Coherent Lightwave Communications Technology

TELECOMMUNICATIONS TECHNOLOGY AND APPLICATIONS SERIES

Series editor:
Stuart Sharrock

The impact of new technologies and new regulatory environments on telecommunications is profound. A convergent communications industry is emerging, embracing computer, networking and software technologies, and including the broadcast, entertainment and consumer electronics industries. The challenge for the future is the design of applications meeting consumer and business demand. The *Telecommunications Technology and Applications Series* focuses on the key new technologies and applications in an environment which crosses many sectors of the communications industry.

Titles available

1. **Coherent Lightwave Communications Technology**
   Edited by Sadakuni Shimda

2. **Network Management**
   Concepts and tools
   Edited by ARPEGE Group

Coherent Lightwave Communications Technology

Edited by
Sadakuni Shimada PhD

Board Director
Fujitsu Laboratories Ltd
Japan

CHAPMAN & HALL
London · Glasgow · Weinheim · New York · Tokyo · Melbourne · Madras

Published by Chapman & Hall, 2–6 Boundary Row, London SE1 8HN, UK

Chapman & Hall, 2–6 Boundary Row, London SE1 8HN, UK

Blackie Academic & Professional, Wester Cleddens Road, Bishopbriggs, Glasgow G64 2NZ, UK

Chapman & Hall GmbH, Pappelallee 3, 69469 Weinheim, Germany

Chapman & Hall Inc., One Penn Plaza, 41st Floor, New York, NY 10119, USA

Chapman & Hall Japan, ITP-Japan, Kyowa Building, 3F, 2-2-1 Hirakawacho, Chiyoda-ku, Tokyo 102, Japan

Chapman & Hall Australia, Thomas Nelson Australia, 102 Dodds Street, South Melbourne, Victoria 3205, Australia

Chapman & Hall India, R. Seshadri, 32 Second Main Road, CIT East, Madras 600 035, India

First edition 1995

© 1995 Sadakuni Shimada

Typeset in 10/12pt Times by Thomson Press (India) Ltd, New Delhi
Printed in Great Britain by St Edmundsbury Press, Bury St Edmunds, Suffolk

ISBN 0 412 57940 5 (HB)

Apart from any fair dealing for the purposes of research or private study, or criticism or review, as permitted under the UK Copyright Designs and Patents Act, 1988, this publication may not be reproduced, stored, or transmitted, in any form or by any means, without the prior permission in writing of the publishers, or in the case of reprographic reproduction only in accordance with the terms of the licences issued by the Copyright Licensing Agency in the UK, or in accordance with the terms of licences issued by the appropriate Reproduction Rights Organization outside the UK. Enquiries concerning reproduction outside the terms stated here should be sent to the publishers at the London address printed on this page.
     The publisher makes no representation, express or implied, with regard to the accuracy of the information contained in this book and cannot accept any legal responsibility or liability for any errors or omissions that may be made.

A catalogue record for this book is available from the British Library

Library of Congress Catalog Card Number: 94 72019

∞ Printed on acid-free text paper, manufactured in accordance with ANSI/NISO Z 39.48–1992 (Permanence of paper).

Contents

List of contributors		ix
Preface		xi
Acknowledgements		xiii

1 Introduction to coherent lightwave communications   1
   (*Kiyoshi Nosu*)

  1.1 Coherent light   1
  1.2 Historical aspects   5
  1.3 The significance of coherent lightwave communication technologies on future telecommunication networks   8

2 Theory of optical coherent detection   13
   (*Katsushi Iwashita*)

  2.1 Optical heterodyne detection   13
     2.1.1 ASK (OOK) system   16
     2.1.2 FSK system   19
     2.1.3 PSK system   21
     2.1.4 Differential detection system   22
  2.2 Optical homodyne detection   23
  2.3 Linewidth influence   24
     2.3.1 Envelope detection   24
     2.3.2 Differential detection   26
     2.3.3 Synchronous detection (PSK heterodyne and homodyne detection)   31
  2.4 Receiver sensitivity comparison   35
  2.5 Power spectrum   39

3 Coherent transmission technologies   43

  3.1 Optical source (*Katsushi Iwashita*)   43
  3.2 Modulators (*Katsushi Iwashita*)   45
     3.2.1 Phase modulator   46
     3.2.2 Direct frequency modulation of laser diodes   47
  3.3 Fiber dispersion (*Nori Shibata*)   49
     3.3.1 Chromatic dispersion   49

	3.3.2	Polarization dispersion	54
3.4		Fiber nonlinearities (*Nori Shibata*)	58
	3.4.1	Stimulated scattering processes	59
	3.4.2	Optical Kerr effect-based phenomena	65
3.5		Heterodyne receiver (*Katsushi Iwashita*)	74
	3.5.1	Thermal noise	74
	3.5.2	Balanced receiver	76
	3.5.3	Phase diversity	79
	3.5.4	Image rejection mixer	82
3.6		Transmission characteristics and delay equalization (*Katsushi Iwashita*)	83
	3.6.1	Transmission characteristics	83
	3.6.2	Delay equalization	83
3.7		Polarization compensation (*Katsushi Iwashita*)	88
	3.7.1	Polarization control	89
	3.7.2	Polarization diversity	91
3.8		Homodyne detection (*Katsushi Iwashita*)	92
	3.8.1	PLL propagation time effect	92

4 Optical filters and couplers — 103

4.1		Optical filters (*Hiromu Toba*)	103
	4.1.1	Configuration of optical filters	103
	4.1.2	Design of waveguide-type optical filter	106
	4.1.3	Tunable optical filters	109
4.2		Mach-Zehnder interferometer waveguide-type filters (*Masao Kawachi*)	113
	4.2.1	Basic interferometer structure	113
	4.2.2	Silica waveguides	114
	4.2.3	Fundamental characteristics	117
	4.2.4	Integration of multiple interferometer units	118
4.3		Optical couplers (*Masao Kawachi*)	120
	4.3.1	Fiber-optic couplers	121
	4.3.2	Integrated-optic waveguides couplers	122

5 Optical frequency division multiplexing systems — 129

5.1		Optical frequency stabilization and measurement (*Osamu Ishida*)	129
	5.1.1	Laser frequency fluctuations	129
	5.1.2	Laser frequency stabilization	131
	5.1.3	Laser frequency measurement	136
5.2		Multichannel frequency stabilization (*Hiromu Toba*)	138
	5.2.1	Requirements	139

		5.2.2 Multichannel frequency stabilization method	140
	5.3	Fiber nonlinear effects in optical FDM systems (*Nori Shibata*)	144
		5.3.1 FDM transmission system configuration	146
		5.3.2 SRS-induced crosstalk	148
		5.3.3 XPM-induced crosstalk	149
		5.3.4 FWM-induced crosstalk	150
	5.4	Channel selective receiver utilizing heterodyne detection (*Hiromu Toba*)	156
		5.4.1 Principle	157
		5.4.2 Channel spacing reduction by the image rejection receiver	162
		5.4.3 Tunable lasers for heterodyne detection	164
		5.4.4 Channel selection technique	165
	5.5	Channel selective receiver utilizing optical filter and direct detection (*Hiromu Toba*)	168
		5.5.1 Principle	169
		5.5.2 Design considerations	172
		5.5.3 Transmission characteristics	180
6	**Optical amplifiers for coherent transmission and optical FDM**		**189**
	6.1	In-line amplifier systems (*Takeshi Ito*)	189
		6.1.1 System performance and amplifier performance	190
		6.1.2 Limitation factors	192
		6.1.3 Fiber dispersion limit	193
		6.1.4 Noise accumulation limit	194
		6.1.5 Nonlinear effect limit	197
		6.1.6 Expected system performance	198
		6.1.7 Measured error rate performance	199
		6.1.8 Factors affecting transmission performance	202
	6.2	Common amplifiers (*Kyo Inoue*)	203
		6.2.1 LD amplifier	203
		6.2.2 Fiber amplifiers	208
		6.2.3 System design	212
7	**Systems applications**		**217**
	7.1	Long-haul trunks and submarine communications (*Takeshi Ito*)	217
		7.1.1 Technologies for long-span transmission systems	217
		7.1.2 Transmission distance of repeaterless system	220
		7.1.3 Outline of experimental repeaterless transmission system	221
		7.1.4 Measured polarization fluctuation and error rate variation	225
		7.1.5 Linewidth and wavelength tuning range of aged DFB-LDs	228

	7.2 Optical FDM networks and switching systems *(Kiyoshi Nosu)*	230
	7.3 Optical SDM and TDM networks and switching *(Kiyoshi Nosu)*	237
8	**Epilogue** *(Kiyoshi Nosu)*	**241**
Index		**251**

List of contributors

Kiyoshi Nosu, NTT Transmission Systems Laboratories
Katsushi Iwashita, NTT Transmission Systems Laboratories
Nori Shibata, NTT Transmission Systems Laboratories
Masao Kawachi, NTT Opto-electronics Laboratories
Hiromu Toba, NTT Transmission Systems Laboratories
Osamu Ishida, NTT Transmission Systems Laboratories
Takeshi Ito, Chiba Institute of Technology
Kyo Inoue, NTT Transmission Systems Laboratories

Preface

This book covers a wide range of technical issues relating to lightwave technologies using high coherence lightwaves. Electromagnetic wave communication started when the first wireless system was invented by Marconi in 1895. However, we had to wait about one hundred years to realize a similar technology in the lightwave frequency region. The invention of lasers in 1960 and two technology innovations in 1970 – low loss silica fiber and semiconductor lasers operating at room temperature – promoted the development of fiber-optic transmission systems. The deployment of high-speed long-haul fiber-optic transmission systems has led to the formation of domestic and international trunk networks. The installed fiber cables in local loop plants provide multimedia communication services including broadband video. However, present lightwave communication systems do not fully utilize the fruitful potential of lightwaves, namely the capacity of extremely high frequency electromagnetic information carrier waves.

The frequency of lightwaves used for fiber-optic transmission is about 200 THz (2×10^{14} Hz), and the frequency bandwidth of the fiber low loss region is about 20 THz (2×10^{13} Hz). Recent developments of narrow spectrum width semiconductor laser and planar optical waveguide devices offer us the possibilities for a new generation of lightwave-based communication systems.

This book focuses on system aspects of the new generation lightwave communication technologies such as optical frequency division multiplexing and coherent detection. Chapter 1 overviews lightwave communication system technology. The status of lightwave communication is compared with those of radio and microwave communication systems. Chapter 2 describes the theory of coherent optical modulation and demodulation to prepare for the discussion on coherent transmission system elements and their system limiting factors in Chapter 3. Chapter 3 discusses elements of coherent transmission systems such as modulators, fiber, heterodyne and homodyne receivers, and delay equalizers.

Chapter 4 describes optical filters and couplers, which are significant in optical frequency division multiplexing systems. Chapter 5 discusses element technologies of optical frequency division multiplexing systems including optical frequency measurement, optical frequency stabilization, fiber nonlinear effect evaluation and receivers for optical FDM systems. Optical amplifiers are discussed in Chapter 6.

The last two chapters describe systems applications. Chapter 7 discusses the possible systems applications such as long-haul trunks, submarine communications, subscriber loops, LAN/MAN, and optical switching. Chapter 8 summarizes the next generation of lightwave technologies and outlines future perspectives on lightwave-based telecommunication networks.

Acknowledgements

I wish to express my sincere thanks to Dr. Hideki Ishio, Executive Manager of Lightwave Communication Laboratory, NTT Transmission Systems Laboratories, for participating in the creation of this book with his thoughtful advice.

All chapters are written by researchers active within NTT Transmission Systems Laboratories, where I conducted research and development of fiber-optic transmission systems from the middle of the 1970s to the beginning of the 1990s.

Sadakuni Shimada
Kawasaki, Japan
September, 1993

1
Introduction to coherent lightwave communications

Kiyoshi Nosu

1.1 COHERENT LIGHT

The study of coherent lightwave communication (Yariv, 1978, Sargent III, *et al.*, 1974, Karaschoff, 1978) must begin by examining the nature of both coherency and lightwaves. What are lightwaves and how can they be coherent? Light is one form of electromagnetic wave. Its significant characteristic is its very high frequency. Figure 1.1 plots part of the spectrum of electromagnetic waves and the applications for communication systems. Commercial radio communication waves range from several kilohertz (kHz) to several tens of GHz. The frequency bands of lightwaves used for optical communication lie between 200 to 300 terahertz (THz). Therefore, in terms of classical physics, light is an electromagnetic wave.

A spectrum of electromagnetic radiation depends upon how the radiation was generated. Radio band waves are generated by the bulk movement of electrons through semiconductors or conductive materials. A lightwave, on the other hand, is generated when an electron falls to a lower energy level. The frequency of the lightwave depends on the amount of energy the electron loses; larger losses produce higher frequencies.

Because lightwaves are generated by discrete atomic events, their duration is extremely brief and coherency is extremely difficult to attain. Existing lightwave sources are able to produce streams of coherent lightwaves only over short periods of time, the **coherent time** of the source. Therefore, lightwaves are considered as a sum of dumped oscillation waves radiated from atoms or molecules.

For example, a black body source can produce such a stream for only 10^{-15} s. A mercury vapour lamp is capable of producing coherent light for 10^{-10}–10^{-11} s, which is equivalent to 10 000–1000 oscillations. Therefore, in general the correlation between light sources is very small. However, the continuous oscillation time of a highly stabilized laser is 10^{-6}–10^{-7} s.

The time of a continuous oscillation is called the coherent time. From the coherent time, we derive the **coherent length** of the light source, another expression

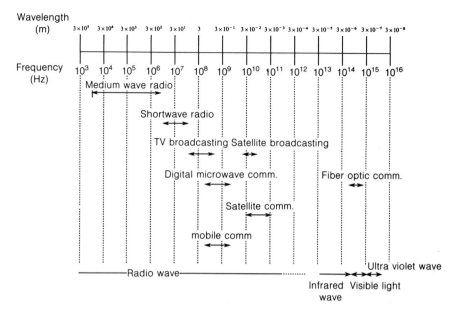

Fig. 1.1 Electromagnetic frequency spectrum.

of the source's ability to produce coherent light. In terms of optical coherent communications systems, we need light sources that have long coherent times and coherent length.

An ideal source of an electromagnetic wave generates infinitely long, continuous wave trains in phase with the frequency and the amplitude constant at all points. Coherency is difficult to achieve because there are so many factors that promote incoherency. The first, multiple lightwaves are generated by multiple point sources which have different spatial locations. This leads to **spatial incoherency**, where the lightwaves are not spatially in phase. The second type is **temporal coherence** which we derived above as coherent time. The finite length of a wave train, temporal amplitude change and frequency fluctuation of lightwaves constitute **temporal incoherency**. This second type of coherency, temporal coherency, is the major source of performance degradation in single mode optical fiber communications systems.

Figure 1.2 shows schematic pictures of coherent and incoherent light. Figure 1.3 shows the spectra of various light sources. In general, light obtained from natural phenomena has a broad spectrum width because it includes the radiation from electrons orbiting various types of atoms. Sunlight, Fig. 1.3(a), is a typical example. The light from an arc lamp, shown in Fig. 1.3(b), also has a broad spectrum, while a more sophisticated lamp, such as a halogen lamp, has strong peaks on its emission spectrum as shown in Fig. 1.3(c). The ideal laser light, which is almost totally coherent light, has a single strong peak.

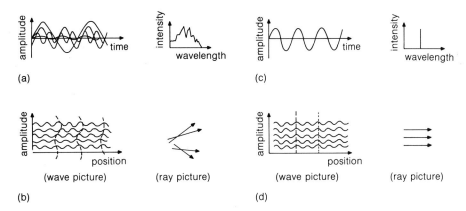

Fig. 1.2 Coherent and incoherent light: (a) temporal incoherency; (b) spatial incoherency; (c), (d) coherent light.

In order to clarify the characteristics of coherent light, we are going to discuss coherence in mathematical terms. The incoherent light model shown in Fig. 1.2 consists of many frequency components. However, in order to simplify the following argument, we will consider only incoherence caused by damped oscillations. Here, a lightwave is considered as the sum of many damped oscillating waves. Therefore, we will start from the damped oscillation characterized by the **damping coefficient**, γ.

A lightwave, then, is expressed as follows.

$$a(t) = A_0 \exp[-(\gamma + i\omega_0)t] \qquad (1.1)$$

We will obtain the spectrum of $a(t)$ as a Fourier transform of $a(t)$:

$$A(\omega) = \int_0^\infty a(t)\exp(-i\omega t)dt$$

$$= \frac{A_0}{[-\gamma + i(\omega_0 - \omega)]}$$

Then, we obtain:

$$|A(\omega)|^2 = \frac{A_0^2}{[\Gamma^2 + (\Omega_0 - \Omega)^2]}$$

The above distribution is a Lorentzian distribution. The half value width of the spectrum, $2\Delta\omega$, is given as follows:

$$2\Delta\omega = 2\gamma$$

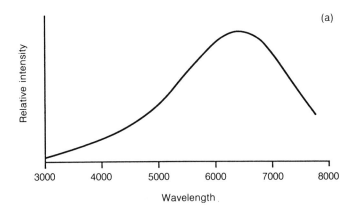
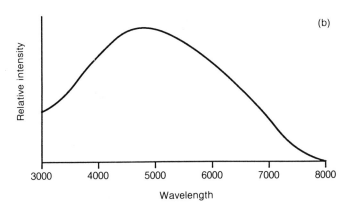
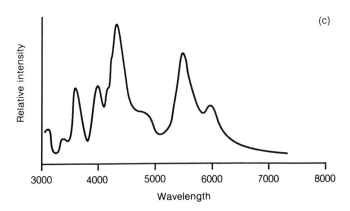

Fig. 1.3 Spectra of various light sources: (a) sunlight; (b) arc lamp; (c) halogen lamp.

HISTORICAL ASPECTS 5

If we define the life time of photon ΔT as the time after which its intensity $|a(t)|^2$ decreases to $1/e$, we obtain $\Delta\omega\,\Delta T = \frac{1}{2}$.

The above equation indicates that the lightwave having a finite life time, ΔT, has a spectral linewidth proportional to $1/\Delta T$. Here the finite life time, ΔT, is the coherent time, described at the beginning of this chapter, which indicates the degree of temporal coherence. Finally, coherence is discussed from the viewpoint of communication theory. In designing communication systems, the evaluation of transmitted signal forms is critical. For example, noise imposed on an optical signal would cause information distortion in analog signals and bit errors in digital signals. Therefore, the finite spectrum linewidth, $\Delta\omega$, of a light source in an optical transmitter results in a signal hazard at the receiver, which depends on modulation and demodulation formats, transmission speed, received power level, etc. The details will be discussed in the following chapters.

1.2 HISTORICAL ASPECTS

Ongoing research into the use of electromagnetic waves as a communication medium has led to the communication revolution. Guglielmo Marconi opened the first gate in 1890 with the invention of wireless communication. Since then, many sophisticated technologies have been developed for wireless communications. The development of radio wave generators occurred well before that of more sophisticated wireless receivers – the history of radio wave oscillators is summarized in Table 1.1.

The arc discharge type generators of the 1890s were used to transmit Morse code signals. In other words, in the early stage of radio, communication was carried out by 'incoherent' communication techniques. After the appearance of vacuum tube oscillators at the beginning of the 20th Century, frequency modula-

Table 1.1 Electromagnetic wave oscillator history

Year	Innovation
1890s	Spark discharge oscillator (damped oscillation wave)
1900s	Arc oscillator (continuous wave)
	High frequency generator (high power continuous wave)
1910s	Vacuum tube oscillator (high frequency wave from a vacuum tube)
1920s	Crystal oscillator (frequency stabilization)
1930s	GT, MT tubes (higher frequency wave)
1940s	Transistor (solid state circuit)
1950s	Maser (stimulated emission radiation)
1960s	Laser (coherent light source)
1970s	Semiconductor laser diode (compact optical transmitter)
1980s	Single mode semiconductor laser (coherent wave transmitter)

Table 1.2 History of electromagnetic wave communication systems

Year	Innovation
1864	Theory of electromagnetic waves (Maxwell)
1888	Electromagnetic wave generation (Hertz)
1895	Wireless communications (Marconi)
1905	Heterodyne detection (Fessenden)
1920	Radio broadcasting (USA)
1928	Frequency modulation (Armstrong)
1960	Laser (Townes)
1967	Optical heterodyne detection experiment with lens waveguides
1970	Low loss optical fiber
	Room temperature operation of semiconductor laser diode
1980–1981	Small/medium capacity fiber optic transmission systems with direct detection
1988	Gbit/s transmission systems with direct detection

tion, heterodyne detection and other advanced technologies were invented so that 'coherent' radio communication technologies have gradually become more common.

In 1960 the first coherent lightwave oscillator was invented (Table 1.2). This was named the Light Amplifier by Stimulated Emission of Radiation, or laser. Immediately after the invention of the laser, much effort was devoted to using the mature (coherent) radio communication technologies developed for the radio frequency band in the lightwave and infrared bands. During the ensuing ten years, the concepts of optical heterodyne detection, optical frequency division multiplexing and integrated optics were proposed. For example, in 1967 AT&T carried out an optical heterodyne detection experiment using an He–Ne gas laser and lens beam guides, as shown in Fig. 1.4.

Early researchers faced intrinsic difficulties whenever attempts were made to introduce elegant radio communication technologies into lightwave communication systems. Some difficulties were as a result of the fact that lightwave frequencies are very high and the wavelengths are very short, namely about 1 μm. For example, an electron cannot match the speed of oscillation of a lightwave of frequency around 200 THz, although a radio wave antenna directly receives radio wave oscillations. Then, the sophisticated receiver configurations developed for radio communications had to be modified. The oscillator frequency stabilization is also more difficult than in radio band oscillators. The absolute carrier frequency stabilization of lightwave systems must be high in proportion to the carrier wave frequency. Therefore, the relative frequency stabilization for lightwave systems would be 10^{-4}–10^{-5} times more severe than the frequency stabilization required for radio systems.

The fact that the wavelength is very short causes difficulties in optical circuit

HISTORICAL ASPECTS

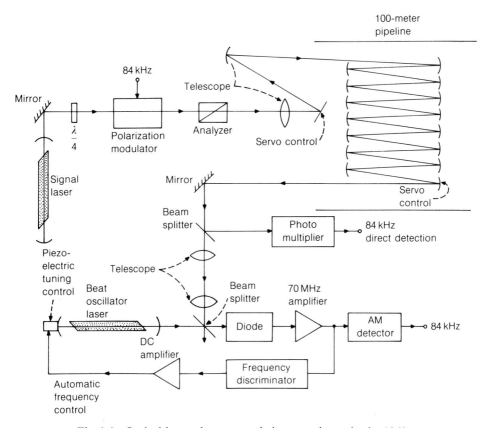

Fig. 1.4 Optical heterodyne transmission experiment in the 1960s.

fabrication. The wavelength of a lightwave is 0.8–1.5 μm, while the radio wave wavelength is longer than 1 m. Therefore, the requirements for the waveguide flatness and phase control accuracy are completely different. This book will describe how we are overcoming these fundamental problems in advanced lightwave communication systems.

In spite of these difficulties, research on fiber optic transmission systems accelerated after the development of low loss silica fiber and with the advent of semiconductor lasers operating at room temperatures in the 1970s. However, current fiber optic communication system technologies only make use of a narrow bandwidth of lightwaves and information is sent by simply changing light source energy. Namely, the communication technology of the current fiber optic communications is at the level of the era of Marconi's technology, where damped oscillation waves (incoherent waves) from an arc discharge generator were used. Therefore, research on coherent lightwave communication was almost abandoned in the 1970s.

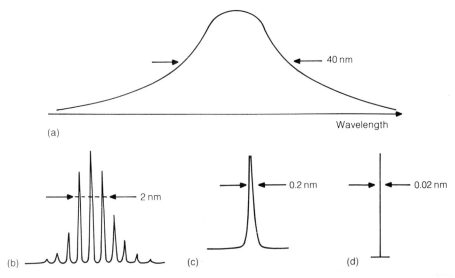

Fig. 1.5 Laser source spectrum widths in modulation: (a) light-emitting diode (LED) for 6.3 Mbit/s transmission (intensity modulation – indirect detection); (b) laser diode (LD) for 400 Mbit/s transmission (intensity modulation – direct detection); (c) LD for 1.6 Gbit/s transmission (intensity modulation – direct detection); (d) LD for 2.4 Gbit/s transmission (CPFSK modulation – heterodyne detection).

In the course of the high speed system development, narrow spectrum width semiconductor lasers, namely coherent lasers, were sought in order to break through the optical fiber chromatic dispersion limitation.

Figure 1.5 shows the emission spectra of various semiconductor light sources. Today, in conventional direct detection systems, DFB laser diodes having 'fair' coherence have been introduced for high speed transmission. This development, namely the use of coherent lasers for incoherent transmission systems, has revived and stimulated research on coherent transmission technologies. Details of this will be described in the following sections and can be found in Okoshi (1979), Yamamoto and Kimura (1981) and Nosu and Iwashita (1988).

1.3 THE SIGNIFICANCE OF COHERENT LIGHTWAVE COMMUNICATION TECHNOLOGIES ON FUTURE TELECOMMUNICATION NETWORKS

Before we discuss the significance of coherent lightwave technologies on future telecommunication networks, we have to define what 'coherent lightwave communication technologies' are. Some readers may consider them to be communication technologies employing coherent light detection. However, the scope of this book is wider than this definition. This book deals with optical communication

FUTURE OF COHERENT LIGHTWAVE TECHNOLOGIES 9

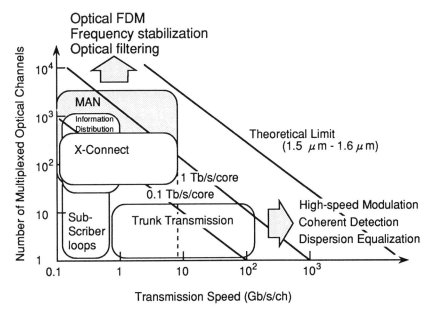

Fig. 1.6 Possible applications for coherent lightwave communication. (MAN – Metropolitan Area Network.)

technologies using coherent lightwaves, namely lightwaves with high coherency. Obviously this definition covers optical coherent detection (optical heterodyne detection and homodyne detection). But it also includes optical frequency division multiplexing (optical FDM), or dense wavelength division multiplexing (dense WDM), which employs coherent lightwaves whose frequencies are well controlled.

Figure 1.6 shows the potential applications of coherent lightwave communication technologies. Coherent detection and optical FDM are two key technologies for coherent lightwave communication systems, although the two technologies are not necessarily implemented simultaneously. For example, a single optical channel point-to-point transmission system employing optical coherent detection, as well as multichannel optical FDM transmission systems employing direction detection, will find applications in future telecommunication networks. For any of these applications, the controllability of the various attributes of coherent lightwaves is essential.

The role of coherent detection is to expand transmission distance as well as transmission capacity in single optical channel transmission, as well as in multichannel optical FDM transmission. The merits of optical coherent detection in future telecommunication networks are:

- high sensitivity;
- electrical signal processing after optical detection;
- channel selectivity.

An Er-doped fiber optical preamplifier can yield sensitivities equivalent to those of coherent detection. However, a theoretical prediction shows that homodyne detection, which is a type of coherent detection, would achieve the highest sensitivity. Electrical signal processing after optical coherent detection realizes compensation against optical fiber chromatic dispersion which restricts transmission distance as well as transmission speed. The channel selectivity of coherent detection in optical FDM will realize high frequency **utilization efficiency**, the ratio of transmission capacity to utilized lightwave frequency bandwidth. The channel selectivity can be the basis for a channel selector in an information distribution system. The coherent detection and the related technologies are discussed in Chapters 2, 3, 6 and 7.

The basic role of optical FDM is to provide an alternative multiplexing scheme to electric time-division multiplexing (TDM), the basis of current digital networks. The merits of optical FDM in future telecommunication networks are:

1. increased channel capacity;
2. flexible optical multiplexing scheme.

Optical FDM can increase transmission capacity per fiber and break through the limit of the functions of current telecommunication nodes based upon electric multiplexing. The most fruitful application of optical FDM would be an optical FDM-based network, which is still at an early research level. As equivalents to TDM-based network elements, optical FDM-based devices such as switches, cross-connects, add/drop multiplexers, and selectors are now being investigated. When these optical FDM-based network elements become feasible, flexible optical networks, which are independent of transmission speed, clock synchronization, and other network node interface restrictions, will be realized. Optical FDM and the related technologies will be discussed in Chapters 4, 5, 6 and 7.

REFERENCES

Delange, O. E. (1968) Optical heterodyne detection, *IEEE Spectrum*, pp. 77–85, Oct.
Delange, O. E. (1968) Optical heterodyne experiments with enclosed transmission paths, *Bell System Technical Journal*, pp. 161–178, Feb.
Garrett, I. (1983) Towards the fundamental limit of optical-fiber communications, *IEEE Journal Lightwave Technology*, **LT–1/1**, pp. 131–138, Feb.
Ishio, H. (1992) Next generation lightwave communication technologies, *NTT Review*, **4/6**, pp. 62–68, Nov.
Karaschoff, P. (1978) *Frequency and Time*, Academic Press.
Nosu, K. (1992) Optical frequency division multiplexing technology and its application to future transmission network, *NTT Review*, **4/6**, pp. 77–82, Nov.
Nosu, K. and Iwashita, K. (1988) A consideration of factors affecting future coherent lightwave communication systems, *IEEE Journal Lightwave Technology*, **LT–5/6**, pp. 686–694, May.

REFERENCES

Okoshi, T. (1979) Feasibility study of frequency-division multiplexing optical fiber communication system using heterodyne or homodyne schemes, paper, Technical Group, IECE Japan, No. OQE78–139, Feb. 27, in Japanese.

Sargent III, M., Scully, M. O., Lamb Jr, W. E. (1974) *Laser Physics*, Addison-Wesley.

Yamamoto, Y. and Kimura, T. (1981) Coherent optical fiber transmission systems, *IEEE Journal of Quantum Electronics*, **17**, pp. 919–934.

Yariv, A. (1978) *Quantum Electronics* (2nd ed), John Wiley & Sons.

2
Theory of optical coherent detection

Katsushi Iwashita

Coherent lightwave communications utilize optical amplitude, phase, and frequency as information-bearing signals. Coherent optical detection, which means optical heterodyne and homodyne detection, offers receiver sensitivity improvement (Yamamato, 1980 and Okashi, *et al.*, 1981) and frequency selectivity improvement. The first advantage will significantly affect long repeater span transmission systems and the second advantage will affect optical frequency division multiplexing (FDM) systems. These features provide a means for exploiting the vast bandwidth of single mode fiber.

In this chapter, bit error rates for several coherent optical detection systems are described, and the performances compared with each other. The influence of spectral linewidth on receiver sensitivity is discussed for several coherent lightwave communication systems.

Here, we define the terms 'coherent optical detection' and 'demodulation' to avoid confusion:

Coherent optical detection. Detection scheme employing nonlinear mixing between two lightwaves. Typically one of these is a signal, the other is a local oscillator wave, and the mixing is performed using a photodetector.

Demodulation. Recovery of baseband signal from the electrical signal.

2.1 OPTICAL HETERODYNE DETECTION

The basic configuration of optical heterodyne detection is shown in Fig. 2.1. The local oscillator frequency is higher or lower than signal light frequency. The optical signal light is combined with local oscillator light and down converted by the photodetector into an electrical signal. The electrical signal is filtered to reject excess noise by a bandpass filter (BPF) and demodulated by a demodulator. The low pass filter (LPF) is used to remove the double frequency component which is generated by the multiplying process.

The electrical field of the optical signal E_S is expressed as follows.

$$E_S(t) = \sqrt{2P_S(t)} \cos\{\omega_S(t)t + \phi_S + \varphi(t)\} \qquad (2.1)$$

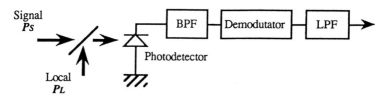

Fig. 2.1 Optical heterodyne detection system configuration.

where $P_S(t)$ is the signal **power**, $\omega_S(t)$ is the signal angular **frequency**, ϕ_S is the **phase** including noise, and $\varphi(t)$ is the modulation phase.

Amplitude, frequency, and phase can be utilized to transmit signal information. If the amplitude is utilized, $P_S(t)$ is modulated by the signal. This modulation is called amplitude shift keying (ASK) or binary data is called on-off keying (OOK). If the frequency and phase are utilized to transmit the data, $\omega_S(t)$, $\varphi(t)$ is modulated by the information signal, respectively. The frequency modulation and phase modulation techniques are called frequency shift keying (FSK) and phase shift keying (PSK), respectively.

The electrical field of the optical signal is illustrated in Fig. 2.2.

The local laser electrical field E_L is expressed as follows.

$$E_L = \sqrt{2P_L} \cos(\omega_L t + \phi_L) \tag{2.2}$$

where P_L is the local oscillator optical power, ω_L is the local oscillator angular frequency, ϕ_L is the phase noise.

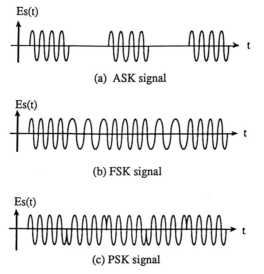

Fig. 2.2 Electrical field of optical signal: (a) ASK signal; (b) FSK signal; (c) PSK signal.

OPTICAL HETERODYNE DETECTION

The signal and local oscillator light are combined and detected by a photodetector. This combined process is assumed to be ideal; the combining method is discussed in more detail in Chapter 3.

The photodetector current $i(t)$ is expressed as follows.

$$i(t) = \frac{\eta e}{h\nu}[P_S + P_L + 2\sqrt{P_S P_L}\{\cos((\omega_S - \omega_L)t + \phi_S - \phi_L + \varphi(t))\}] \quad (2.3)$$

where η is the quantum efficiency of the photodetector, e is the electron charge, h is Planck's constant (6.625×10^{-34} J s), ν is the optical frequency.

The first and second terms are the photo currents resulting from local power and signal power, respectively. The third term is the beat signal which is proportional to the signal electrical field and the local electrical field. The waveform in this process is shown schematically in Fig. 2.3.

The electrical signal power S and shot noise power N_S are expressed as follows.

$$S = \frac{\left(\frac{2\eta e}{h\nu}\sqrt{P_S P_L}\right)^2}{2} = 2R^2 P_S P_L \quad (2.4)$$

$$N_S = 2e\frac{\eta e}{h\nu}(P_L + P_S)B = 2eR(P_L + P_S)B \quad (2.5)$$

where

$$R = \frac{\eta e}{h\nu}$$

and B is receiver bandwidth.

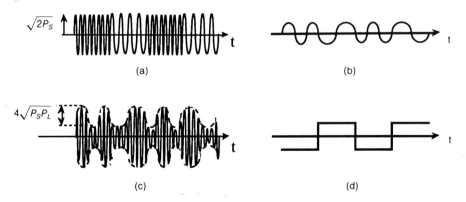

Fig. 2.3 Waveform at heterodyne detection process: (a) optical signal with FSK modulation; (b) optical signal combination with local oscillator light; (c) IF electrical signal; (d) demodulated signal.

Therefore, the **signal-to-noise ratio** including electrical noise N_{th} is expressed as follows.

$$S/N = \frac{\frac{\{2R\sqrt{P_S P_L}\}^2}{2}}{\{2eR(P_S + P_L)B + N_{th}\}}$$

$$= \frac{2R^2 P_S P_L}{2eR(P_S + P_L)B + N_{th}} \tag{2.6}$$

Since the amplitude of an intermediate frequency signal is proportional to the product of signal and local oscillator amplitude, the IF power can increase compared with electrical noise. Therefore, coherent optical detection offers improved sensitivity overcoming thermal noise. The sensitivity for coherent optical detection is limited by the quantum noise present in the detected photo current. This noise increases linearly with local oscillator power so that the electrical signal-to-noise ratio approaches a constant value, which is called the **shot noise limit**.

$$S/N = \frac{\frac{\{2R\sqrt{P_S P_L}\}^2}{2}}{\{2eR(P_S + P_L)B + N_{th}\}} \cong \frac{RP_S}{eB} \tag{2.7}$$

If we assume the receiver equivalent input current noise density is $10\,\text{pA}/\sqrt{\text{Hz}}$, $\eta = 0.8$, $\lambda = 1.55\,\mu\text{m}$ and $P_L = 5\,\text{dBm}$, the signal-to-noise ratio is 3.2 dB down from the shot noise limited value. It is very close to the shot noise limit. Therefore, coherent optical detection can realize the shot noise limit fairly easily.

2.1.1 ASK (OOK) system

The ASK system configuration is shown in Fig. 2.4. The transmitted optical signal is combined with the local light. The combined light is converted into an electrical IF signal by a photodetector (PD). In our model, the signal is contaminated at the input with additive Gaussian noise resulting from local oscillator light shot noise

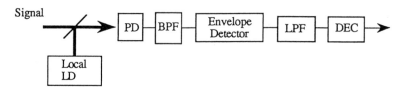

Fig. 2.4 Receiver configuration of ASK system.

OPTICAL HETERODYNE DETECTION

and receiver thermal noise. The signal with noise is applied to an envelope detector and filtered to eliminate excess noise. The demodulated signal is applied to a decision circuit (DEC) to determine the existence of a mark or a space.

The data modulates the amplitude and the detected IF signal is expressed as follows according to Schwartz, Bennett and Stein (1966):

$$r(t) = 2R\sqrt{P_S(t)P_L}\cos\omega_{IF}t + n_x(t)\cos\omega_{IF}t - n_y(t)\sin\omega_{IF}t$$
$$= \sqrt{[2R\sqrt{P_S(t)P_L} + n_x(t)]^2 + n_y(t)^2}\cos(\omega_{IF}t + \phi) \quad (2.8)$$

where

$$\phi = \tan^{-1}\left(\frac{n_y}{2R\sqrt{P_S(t)P_L} + n_x}\right)$$

$\omega_{IF} = \omega_S - \omega_L$ is the intermediate angular frequency, n_x and n_y are noise with narrow band representation.

If we assume that (n_x, n_y) have a Gaussian distribution and are independent of each other with a zero mean value and a variance of σ^2, the probability density function is given by:

$$p(n_x, n_y) = \frac{1}{2\pi\sigma^2}\exp\left[\frac{-(n_x^2 + n_y^2)}{2\sigma^2}\right] \quad (2.9)$$

The probability density function of equation 2.9 with amplitude and phase is expressed as:

$$p(\rho, \phi) = \frac{\rho}{2\pi\sigma^2}\exp\left[\frac{-(\rho^2 + A^2 - 2A\rho\cos\phi)}{2\sigma^2}\right] \quad (2.10)$$

where

$$\rho = \sqrt{[2R\sqrt{P_S(t)P_L} + n_x(t)]^2 + n_y(t)^2}$$
$$A = 2R\sqrt{P_S(t)P_L} \quad (2.11)$$

The probability density function of amplitude can be expressed by integrating equation 2.10 from 0 to 2π:

$$p(\rho) = \frac{\rho}{2\pi\sigma^2}\exp\left[\frac{-(\rho^2 + A^2)}{2\sigma^2}\right]\int_0^{2\pi}\exp\frac{(\rho A\cos\phi)}{\sigma^2}d\phi$$
$$= \frac{\rho}{\sigma^2}I_0\left[\frac{A\rho}{\sigma^2}\right]\exp\left[\frac{-(\rho^2 + A^2)}{2\sigma^2}\right] \quad (2.12)$$

where $I_0(x)$ is the modified Bessel function.

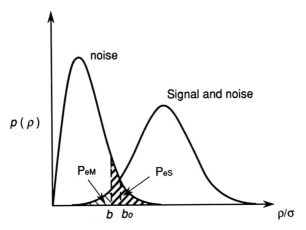

Fig. 2.5 The probability of error.

The probability of error is shown in Fig. 2.5.

To determine if the data is a mark or space, the threshold level b is defined. If the received signal voltage is less than b, the signal is determined to be a space. Conversely if the received voltage is more than b, the signal is determined to be a mark. The error probability is determined by the probability, P_{eM}, at which the received voltage of the transmitted mark falls below b and the probability, P_{eS}, at which the received voltage of the transmitted space exceeds b.

The probability is calculated as follows.

$$P_{eM} = \int_0^b p(r)\,dr$$

$$= 1 - \int_b^\infty \frac{r}{N} I_0\left(\frac{ur}{N}\right) e^{\frac{-(r^2+u^2)}{2N}}\,dr$$

$$= 1 - Q[\sqrt{2\gamma}, b] \qquad (2.13)$$

where $Q[x, y]$ is the Marcum Q function, γ is the signal-to-noise ratio at the output of the bandpass filter:

$$\gamma = \frac{A^2}{2\sigma^2} = \frac{2R^2 P_S P_L}{2eR(P_S + P_L)B + N_{th}} \qquad (2.14)$$

and if a space is transmitted:

$$P_{eS} = \int_b^\infty p(r)\,dr = e^{\frac{-b^2}{2}} \qquad (2.15)$$

If we assume that the mark and space have the same probability, the error probability is expressed as:

$$P_e = \frac{1}{2}P_{eM} + \frac{1}{2}P_{eS} = \frac{1}{2}[1 - Q(\sqrt{2\gamma}, b) + e^{\frac{-b^2}{2}}] \tag{2.16}$$

This error rate depends on the threshold level b which, therefore, should be optimized. This level can be obtained by solving:

$$\frac{\partial P_e}{\partial b} = 0$$

The approximate solution is expressed as follows.

$$b_0 \cong \sqrt{2 + \frac{\gamma}{2}} \tag{2.17}$$

The optimum value depends on the received power. The optimized error rate is given as:

$$P_e \cong \frac{1}{2}e^{\frac{-\gamma}{4}} \tag{2.18}$$

Note that the required signal-to-noise ratio to obtain a 10^{-9} error rate is 19 dB. The required number of photons to detect 1 bit within a 10^{-9} error rate is 80 when $\eta = 1$. These requirements are shown by the peak level of the signal. In terms of the average level, the required signal-to-noise ratio is 16 dB.

ASK systems can be detected using synchronous detection. The error rate is given by:

$$P_e = \frac{1}{2}\text{erfc}\left(\frac{\sqrt{\gamma}}{2}\right) \tag{2.19}$$

The receiver sensitivity difference between the synchronous detection and the envelope detection becomes smaller as the signal-to-noise ratio increases. The required peak and average signal-to-noise ratio is 18.5 dB and 15.5 dB, respectively. The difference is only 0.5 dB at 10^{-9} error rate.

2.1.2 FSK system

FSK is the most popular scheme for optical coherent communication systems because the FSK signal is obtained by simply modulating the current of a laser diode.

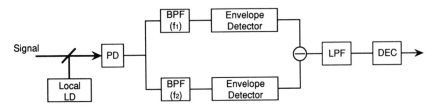

Fig. 2.6 Receiver configuration of FSK system.

The configuration of the FSK receiver is shown in Fig. 2.6. We assume the FSK signal is a frequency-modulated rectangular signal of constant amplitude.

The heterodyned signal (frequency f1 and f2) is passed through two bandpass filters with different pass bands, and then the signals are envelope detected. These detected signals are differentially combined

The error rate is obtained by the following scheme.

If frequency f1 is transmitted, the probability density function of the output voltage ρ_1 of the bandpass filter (BPF) f1 from equation 2.12 is expressed by:

$$p(\rho_1) = \frac{\rho_1}{\sigma^2} I_0 \left[\frac{A\rho_1}{\sigma^2} \right] \exp\left[\frac{-(\rho_1^2 + A^2)}{2\sigma^2} \right] \quad (2.20)$$

The probability density function of bandpass filter f2 with the output voltage ρ_2 is expressed by:

$$p(\rho_2) = \frac{\rho_2}{\sigma^2} \exp\left[\frac{-(\rho_2^2 + A^2)}{2\sigma^2} \right] \quad (2.21)$$

The probability of error is obtained by calculating the probability of $\rho_2 > \rho_1$, and can be expressed as:

$$P_e = \text{prob}(\rho_2 > \rho_1) = \int_{\rho_1=0}^{\infty} p(\rho_1) \left[\int_{\rho_2=\rho_1}^{\infty} p(\rho_2) d\rho_2 \right] d\rho_1 \quad (2.22)$$

The error rate is expressed as:

$$P_e = \frac{1}{2} e^{\frac{-\gamma}{2}} \quad (2.23)$$

where γ is the signal-to-noise ratio which is defined as:

$$\gamma = \frac{A^2}{2\sigma^2} = \frac{2R^2 P_S P_L}{2eR(P_S + P_L)B + N_{th}} \quad (2.23a)$$

OPTICAL HETERODYNE DETECTION   21

The optimum threshold level does not depend on the signal-to-noise ratio while the threshold level of ASK systems does depend on the signal-to-noise ratio. Therefore, setting the threshold level at 0 optimizes the receiver.

Note that the required signal-to-noise ratio to obtain a 10^{-9} error rate is 16 dB. The required number of photons to detect 1 bit within a 10^{-9} error rate is 40 where $\eta = 1$. This is only half the value demanded by ASK. However, this shows peak level; if we compare with the average level, the required ratio is the same as the ASK system.

2.1.3 PSK system

PSK is the most sensitive scheme for optical coherent heterodyne detection. The configuration of the PSK receiver is shown in Fig. 2.7. The IF signal is multiplexed with a local oscillator signal by using a mixer. The double frequency component is eliminated by a low pass filter and applied to the decision circuit. The local oscillator signal has to track the signal phase, which requires a phase locked loop.

The received IF signal is expressed as follows.

$$r(t) = 2R\sqrt{P_S(t)P_L}\cos\{\omega_{IF}t + \varphi(t)\} + n_x(t)\cos\omega_{IF}t - n_y(t)\sin\omega_{IF}t \qquad (2.24)$$

The information is carried by the phase $\varphi(t)$.

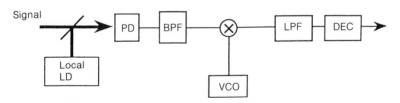

Fig. 2.7 Receiver configuration of PSK system.

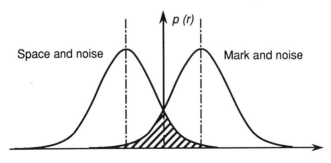

Fig. 2.8 The probability of error.

If the frequency and phase of the voltage control oscillator (VCO) are matched to those of the signal, the received signal can be expressed as follows.

$$r(t) \approx 2R\sqrt{P_S P_L} a_n(t) + n_x(t) \qquad (2.25)$$

where a_n has the value ± 1.

If the mark is transmitted, the received signal $r(t)$ is positive. The probability density function with output voltage $r(t)$ can be described as follows.

$$p(r) = \frac{e^{\frac{-(r-u)^2}{2\sigma^2}}}{\sqrt{2\pi\sigma^2}} \qquad (2.26)$$

The probability of error is shown in Fig. 2.8.

$$\begin{aligned} P_e &= \frac{1}{2}\int_0^\infty p(r)dr \\ &= \frac{1}{2}\int_0^\infty \frac{e^{\frac{-r^2}{2\sigma^2}}}{\sqrt{2\pi\sigma^2}}dr = \frac{1}{2}\text{erfc}(\gamma) \end{aligned} \qquad (2.27)$$

Note that the required signal-to-noise ratio to obtain a 10^{-9} error rate is 18 (12.5 dB). This is 3.5 dB better than FSK.

2.1.4 Differential detection system

Synchronous PSK detection requires a phase locked loop, which complicates receiver circuits. The other method of detecting phase is differential PSK (DPSK) which compares the phase change with the previous bit and differential encoding. For example, when the phase difference between the previous bit is 0°, it assumes a space, and when the phase difference is 180°, it assumes a mark, and the receiver compares the phase between two bits. The configuration of the DPSK receiver is shown in Fig. 2.9.

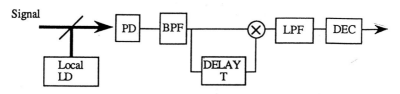

Fig. 2.9 Receiver configuration of DPSK system.

The error rate is expressed as:

$$P_e = \frac{1}{2} e^{-\gamma} \quad (2.28)$$

The required signal-to-noise ratio to get the same error rate is half that of FSK and 0.5 dB lower than that of PSK. This is because the PSK synchronous detection uses a pure oscillator as the reference signal while DPSK uses a pulse which is less noisy than that used in PSK.

2.2 OPTICAL HOMODYNE DETECTION

Optical homodyne detection matches the transmitted signal phase to that of a local oscillator signal. First we will consider the situation of phase matching according to Glance (1986), Fig. 2.10. The transmitted signal is given by:

$$r(t) = 2R\sqrt{P_S P_L} \cos\{\pi a_k s(t)\} \quad (2.29)$$

where a_k has the value ± 1, $s(t)$ is the modulating waveform. This is a baseband signal; therefore, the error rate is the same as that of the baseband system.

The shot noise power induced by local and signal powers is expressed as follows.

$$\sigma_{sh}^2 = 2eR(P_L + P_S) \int_0^\infty |H(j\omega)|^2 d\omega \quad (2.30)$$

where $H(j\omega)$ is the transfer function of the receiver. If the transfer function is a selected matched filter such as:

$$|H(j\omega)|^2 = \left[\frac{\sin(\omega T/2)}{\omega T/2}\right]^2 \quad (2.31)$$

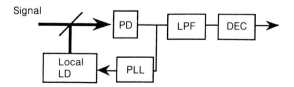

Fig. 2.10. Receiver configuration of optical homodyne detection system.

where T is the bit period time, the noise power becomes:

$$\sigma_{sh}^2 = \frac{eR(P_L + P_S)}{T}$$

$$\approx \frac{eRP_L}{T} \qquad (2.32)$$

where $P_L \gg P_S$ is assumed. The signal-to-noise ratio is:

$$\gamma = \frac{2\eta P_S T}{h\nu} \qquad (2.33)$$

The probability of error is the same as that of synchronous detection:

$$P_e = \frac{1}{2}\text{erfc}[\sqrt{\gamma}] \qquad (2.34)$$

This sensitivity is 3 dB better than that of heterodyne PSK. This is because the required bandwidth for homodyne detection is half that required by heterodyne PSK.

2.3 LINEWIDTH INFLUENCE

Phase or frequency noise in lasers is a well-known phenomenon. It has been observed that the spectral density of this frequency noise has a $1/f$ characteristic up to around 1 MHz, and is flat for frequencies above 1 MHz. The flat or 'white' component is associated with quantum fluctuations and is the principal cause for linewidth broadening. For optical communication sources, the low frequency component can be easily tracked and we can ignore this part of the noise. Therefore, we focus on the white noise component.

Laser diodes typically have spectral linewidths in the range 1 to 50 MHz, which is too broad compared with microwave oscillators which have linewidths on the order of 1 Hz.

Here we will discuss the significant spectral linewidth influence on several coherent optical detection systems.

2.3.1 Envelope detection

Envelope detection measures the power level of the intermediate frequency. If the light sources have significant linewidth, the measured intermediate frequency depends on the signal phase. If we assume that the probability density function of the intermediate frequency deviation resulting from a phase fluctuation is $p_{IF}(\Delta\omega)$, and that of the error in detecting the symbol conditioned on frequency deviation $\Delta\omega$, and receiver power p_S, the total error rate is calculated as follows.

$$P_T = \int_{-\infty}^{\infty} P_C(p_S, \Delta\omega) p_{IF}(\Delta\omega) d(\Delta\omega) \quad (2.35)$$

The probability density function for the frequency deviation $\Delta\omega$ is expressed as Gaussian according to Garrett and Jacobsen (1986):

$$p_{IF}(\Delta\omega) = \frac{1}{\sqrt{\Delta v B T}} \exp\left(\frac{-\Delta\omega^2}{4\pi\Delta v B}\right) \quad (2.36)$$

where Δv is the full IF linewidth at half-maximum of the power spectral density and T is the bit period.

This procedure can be applied to ASK and FSK systems. Here we consider a dual filter FSK system.

Results for error probability P_T were evaluated for a range of IF linewidths and frequency shifts, assuming bandpass filters with full raised-cosine response envelopes (Garrett and Jacobsen, 1986). The calculated results for P_T are shown in Fig. 2.11 with the IF linewidth as a parameter, for different modulation indices m (= frequency deviation/bit rate) and filter bandwidths. For small IF linewidth and relative bandwidth $BT = 1$, the results are close to the shot noise limit for FSK. Systems with $m = 1.5$ are not viable for normalized IF linewidths (ΔvT) much above 0.05, while if $m = 3$, IF linewidths up to at least 0.15 may be tolerated. The effect of increasing the filter bandwidth with a corresponding increase in

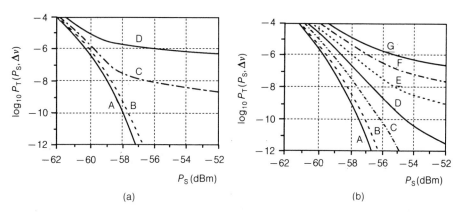

Fig. 2.11 Error probability versus signal power with IF linewidth as parameters for 140 Mbit/s dual filter FSK system: (a) $m = 1.5$, (b) $m = 3.0$. Curve A, $\Delta v = 0.001$; B, 0.05; C, 0.1; D, 0.15; E, 0.2; F, 0.25; G, 0.3.
Source: Garrett and Jacobsen (1986).

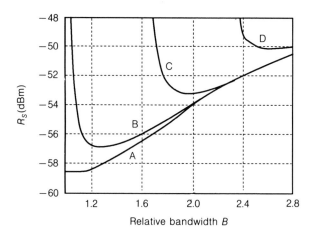

Fig. 2.12 Receiver sensitivity (for 10^{-9} error rate) as a function of relative bandwidth and IF linewidth for 140 Mbit/s dual filter FSK system: curve A, 0.001; B, 0.1; C, 0.2; D, 0.3. Source: Garrett and Jacobsen (1986).

modulation index produces a penalty in the zero linewidth sensitivity because of the increased noise bandwidth and the increased f^2 receiver noise.

The error rate performance depends on the IF bandpass filter bandwidth. Figure 2.12 shows the effect of IF filter bandwidth. Receiver sensitivity is shown as a function of B, with the linewidth as a parameter. The modulation index is given as:

$$m = 2B - 0.5 \qquad (2.37)$$

Namely, the ideal filter is a matched filter with a bandwidth equal to the bit rate. However, if the linewidth is not negligible, a broader bandwidth is optimum. In the case of ASK systems, the error rate floor also depends on the threshold level.

2.3.2 Differential detection

We will consider two types of differential detection schemes: DPSK and CPFSK with differential detection.

(a) DPSK systems

The DPSK receiver configuration is shown in Fig. 2.9. The heterodyned signal is expressed as follows (Nicholson, 1984).

$$r(t) = 2R\sqrt{P_S P_L}\cos(\omega_{IF}t + \phi_S(t)) + n_x(t)\cos\omega_{IF}t - n_y(t)\sin\omega_{IF}t \qquad (2.38)$$

The phase $\phi_S(t)$ is expressed by:

$$\phi_S(t) = \varphi_S(t) + \{\varphi_n(t) - \varphi_n(t+T)\} - \{\varphi_{pS}(t) - \varphi_{pS}(t+T)\} \\ - \{\varphi_{pL}(t) - \varphi_{pL}(t+T)\} \quad (2.39)$$

where the first term is the data which takes values of 0 or π. The second term represents the phase noise due to the shot noise. The third and fourth terms are the quantum phase noise resulting from the transmitter and local oscillator, respectively. The probability of error is given by:

$$P_e = \int_{-\pi/2}^{\pi} \int_{-\infty}^{\infty} p_n(\phi_1 - \phi_2) p_q(\phi_1) d\phi_1 d\phi_2 \quad (2.40)$$

where $p_n()$ is the probability density function PDF of the phase noise due to the shot noise, and $p_q()$ is the PDF of the total quantum phase noise generated by the transmitter and local oscillator.

The probability of error is derived by using equations taken from Blachman (1981):

$$p_n(\phi_1 - \phi_2) = \frac{1}{2\pi} + \frac{\rho e^{-\rho}}{\pi} \sum_{m=1}^{\infty} a_m \cos\{m(\phi_1 - \phi_2)\}$$

$$a_m \sim \left\{ \frac{2^{m-1}\Gamma[(m+1)/2]\Gamma[m/2+1]}{\Gamma[m+1]} \left[I_{(m-1)/2}\left(\frac{\rho}{2}\right) + I_{(m+1)/2}\left(\frac{\rho}{2}\right) \right] \right\}^2 \quad (2.41)$$

where $\Gamma()$ is the gamma function and I_n is the modified Bessel function of the first kind. The PDF of the total quantum phase noise according to Yamamoto and Kimura (1981) is modelled by:

$$p_q(\phi_1) = \frac{1}{\sqrt{2\pi D\tau}} \exp\left(\frac{\phi_1^2}{2D\tau}\right) \quad (2.42)$$

where D is the phase diffusion constant and $\Delta v = \Delta v_R + \Delta v_L = D/2\pi$ is the sum of the transmitter and local oscillator full linewidths at half maximum intensity. Substituting equations 2.41 and 2.42 into 2.40, the probability of error can be expressed as:

$$P_e = \frac{1}{2} - \frac{\rho e^{-\rho}}{2} \sum_{n=0}^{\infty} \frac{(-1)^n}{2n+1} \exp[-(2n+1)^2 \pi \Delta v T] \left[I_n\left(\frac{\rho}{2}\right) + I_{n+1}\left(\frac{\rho}{2}\right) \right]^2 \quad (2.43)$$

This equation gives the probability of error versus received power. Figure 2.13 shows the degradation in optical receiver sensitivity for the probability of error

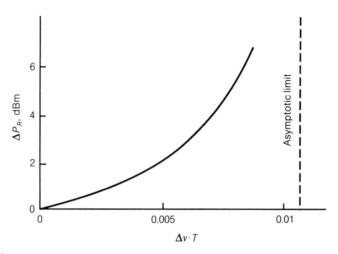

Fig. 2.13 Degradation in optical receiver sensitivity at 10^{-9} for DPSK system. Source: Nicholson (1984).

at 10^{-9} as a function of the product IF linewidth and bit period. Note that the total source linewidth should not exceed 0.33% of the bit rate with a 1 dB power penalty at 10^{-9} bit error rate. The required linewidth depends upon the bit rate because the delay time of the differential detection is $1/B$. For example, in a 1 Gbit/s DPSK system, this criterion gives a total linewidth of less than 3.3 MHz.

(b) CPFSK differential detection

The error probability of differentially detected CPFSK is derived by taking into consideration the delay time of the differential detector, frequency deviation, and phase noise (Iwashita and Matsumoto, 1987). The differential detector configuration and the frequency-to-output voltage conversion relation are shown in Fig. 2.14. The detected signal phase at the shot noise limit in the output of the lowpass filter can be expressed as follows.

$$\phi(t) = \omega_c \tau + a_n \frac{\Delta \omega}{2} \tau + \varphi(t) + \varphi_n(t) \tag{2.44}$$

where $\omega_c = 2f_c = (2n+1)(\pi/2\tau)$, and τ is the differential detection delay time, $\Delta \omega$ is the angular frequency deviation, $\varphi(t)$ is the phase noise due to shot noise, $\varphi_n(t)$ is the phase noise due to the transmitter and local oscillator quantum phase noise, and a_n is the binary data taking values $+1$ or -1.

Integrating with respect to $\phi(t)$ from $-(\Delta \omega/2)\tau$ to $\pi - (\Delta \omega/2)\tau$, we obtain the

LINEWIDTH INFLUENCE

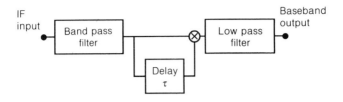

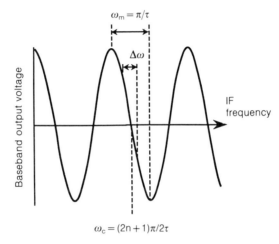

Fig. 2.14 Configuration of CPFSK differential detection and frequency-to-voltage conversion of differential detection.

error probability:

$$P_e = \int_{-\Delta\omega\tau/2}^{\pi - \Delta\omega\tau/2} \int_{-\infty}^{\infty} p_n(\phi_1 - \phi_2) p_q(\phi_1) d\phi_1 d\phi_2 \qquad (2.45)$$

As with the DPSK system, substituting equations 2.41 and 2.42 into 2.45 yields:

$$P_e = \frac{1}{2} \frac{\rho e^{-\rho}}{2} \sum_{n=0}^{\infty} \frac{(-1)^n}{2n+1} \exp[-(2n+1)^2 \pi \Delta v \tau][I_n(\rho/2) + I_{n+1}(\rho/2)]^2$$
$$\times \exp[-(2n+1)^2 \pi \Delta v \tau] \cos\{(2n+1)\alpha\} \qquad (2.46)$$

$$\alpha = \frac{\pi(1-\beta)}{2}$$

$$\beta = \frac{\Delta\omega}{\omega_m} = \frac{2m\tau}{T_0}$$

where ω_m is the available maximum angular frequency deviation, m is the modulation index, and T_0 is the pulse period. The modulation index parameter β is defined as the ratio of the actual frequency deviation to the maximum frequency deviation.

The receiver sensitivity degradation at a 10^{-9} bit error rate with the absence of laser diode (LD) phase noise as a function of the product of beat linewidth and delay time is shown in Fig. 2.15. The required linewidth given a 1 dB power penalty as a function of the modulation index, derived from Fig. 2.15, is shown in Fig. 2.16. The allowable linewidth narrows as the frequency deviation becomes small. The relationship between the required linewidth and the modulation index is:

$$\Delta v T_0 < 6.8 \times 10^{-3} m \qquad (2.47)$$

In the case of m = 0.5, the required linewidth is 0.34% of the bit rate which is the same as for DPSK. The 1 bit delay differential detection of CPFSK is the same as DPSK in its phase noise requirement. However, CPFSK is more flexible than DPSK. When m = 1 and $\tau = T_0/2$, the linewidth requirement is $\Delta v T_0 < 6.8 \times 10^{-3}$.

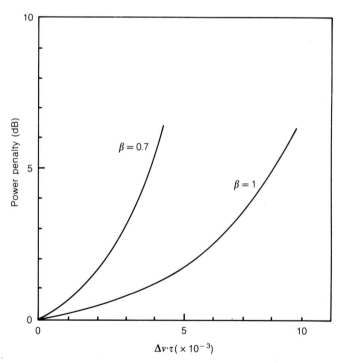

Fig. 2.15 Power penalty from the absence of laser diode phase noise at a 10^{-9} error rate as a function of the product of beat linewidth and delay time.
Source: Iwashita and Matsumoto (1987).

LINEWIDTH INFLUENCE

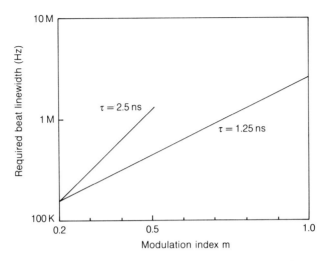

Fig. 2.16 Required beat linewidth given a 1 dB optical power penalty as a function of the modulation index m at 400 Mbit/s modulation.
Source: Iwashita and Matsumoto (1987).

The system becomes twice as tolerant towards the LD phase noise compared with DPSK. This is because the delay time of the differential detection depends upon the modulation index. The delay time for the case of m = 1 is $T_0/2$ and that for m = 0.5 is T_0.

2.3.3 Synchronous detection (PSK heterodyne and homodyne detection)

Synchronous detection schemes, such as PSK heterodyne detection and homodyne detection, will now be discussed. The probability of bit error with steady phase error for PSK heterodyne detection is given by Prabhu (1976) as:

$$P_p = \frac{1}{2}\text{erfc}[\gamma \cos \phi] \qquad (2.48)$$

where ϕ is the phase error. The bit error rate considering the phase error variance is obtained by integrating equation 2.48 while taking into consideration the probability density function of the phase distribution, which is given by:

$$P_e = \int_0^{2\pi} P_p(\phi) p_\phi(\phi) d\phi \qquad (2.49)$$

If we assume the phase error variance distribution is Gaussian, then:

$$p_\phi(\phi) = \frac{1}{\sqrt{2\pi\sigma}} \exp\left[\frac{\phi^2}{2\sigma^2}\right] \tag{2.50}$$

Thus the bit error rate is given by:

$$P_e = \frac{1}{2}\text{erfc}[\rho] + \sum_{l=0}^{\infty}(-1)^l H_l[\rho]\left\{1 - \exp\left[\frac{(2l+1)^2\sigma^2}{2}\right]\right\} \tag{2.51}$$

where $H_l[\rho]$ is defined by:

$$H_l[\rho] \equiv \frac{\rho\exp\left(\dfrac{\rho^2}{2}\right)}{\sqrt{\pi(2l+1)}} \left\{ I_l\left(\frac{\rho^2}{2}\right) + I_{l+1}\left(\frac{\rho^2}{2}\right) \right\} \tag{2.52}$$

If the phase error variance can be obtained, the error rate will be obtained. The error rate performance depends on the phase-locked loop configuration.

Here we will analyze a decision-directed loop which is most tolerable to the phase noise in an optical homodyne detection system. A linearized model of the decision-driven optical phase-locked loop is shown in Fig. 2.17. The closed-loop transfer function of the PLL is given by:

$$H(f) = \frac{G(j2\pi f)^{-1}F(f)\exp(-j2\pi f\tau)}{1 - G(j2\pi f)^{-1}F(f)\exp(-j2\pi f\tau)} \tag{2.53}$$

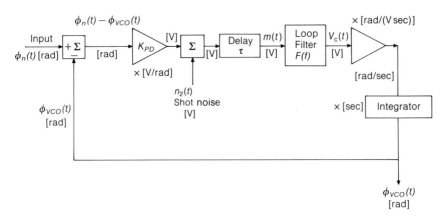

Fig. 2.17 A linearized model of the decision-driven optical phase-locked loop.

where G is the overall loop gain:

$$G = 2G_{\text{VCO}} Rrk\sqrt{P_s P_{\text{LO}}}$$

and where $F(f)$ is the transfer function of the loop filter and τ_p is the propagation delay time of the PLL, r is the detector's load resistance and k is the power splitting ratio of the hybrid.

The phase noise and the shot noise processes are independent of each other; therefore the PLL performance is characterized by the phase-error variance defined by:

$$\sigma^2 \equiv \sigma_{\text{PN}}^2 + \sigma_{\text{SN}}^2 \qquad (2.54)$$

where σ_{PN}^2 represents the phase-error variance due to the phase noise and σ_{SN}^2 represents the phase-error variance due to the shot noise. The phase-error variance due to shot noise can be evaluated thus:

$$\sigma_{\text{SN}}^2 = \frac{S_{\text{SN}}}{K_{\text{PD}}^2} \int_{-\infty}^{\infty} |H(f)|^2 df = \frac{S_{\text{SN}} \cdot B_n}{K_{\text{PD}}^2} \quad (\text{rad}^2) \qquad (2.55)$$

where

$$B_n = \int_{-\infty}^{\infty} |H(f)|^2 df$$

which is the noise bandwidth, and:

$$S_{\text{SN}}(f) = eRP_{\text{LO}} k_{\text{LO}} r^2 \, (\text{V}^2/\text{Hz})$$

is the power spectral density of the shot noise. The phase-error variance due to the phase noise is given by:

$$\sigma_{\text{PN}}^2 = \int_{-\infty}^{\infty} S_{\text{PN}}(f) |1 - H(f)|^2 df = \frac{\delta v}{2\pi} \int_{-\infty}^{\infty} \left| \frac{1 - H(f)}{f} \right|^2 df \quad (\text{rad}^2) \quad (2.56)$$

where

$$S_{\text{PN}}(f) = \frac{\Delta v}{2\pi f^2} \quad (\text{rad}^2/\text{Hz})$$

is the power spectral density of the phase noise, Δv(Hz) is the beat linewidth (FWHM) of the transmitter's light and the local oscillator's light.

As for the loop filter, the first order active filter is adopted and its transfer function $F(f)$ is denoted by:

$$F(f) = \frac{1 + j2\pi f \tau_2}{j2\pi f \tau_1} \qquad (2.57)$$

The total phase-error variance is obtained:

$$\sigma^2 = \frac{e}{2RkP_s} B_n + 2.36\Delta\nu \frac{1}{B_n} \qquad (2.58)$$

Figure 2.18 shows the phase-error standard deviation versus the normalized loop bandwidth for several values of the power-to-linewidth ratio, PLR = $kP_s/\Delta\nu$. For small loop bandwidth, the phase error is dominated by phase noise. As the loop bandwidth is increased, the phase error is decreased. For large loop bandwidths, the phase error is dominated by shot noise. Therefore, as the loop bandwidth is increased, the phase error is increased. Consequently, there is an optimum value of the loop noise bandwidth. This figure also shows that the required power to a reliable phase lock (phase-error variance is less than 10°) is more than 0.8 pW of signal power for every kilohertz of the laser linewidth.

The required spectral linewidth is obtained by assuming that the total power penalty is 1 dB (0.5 dB to finite phase error and 0.5 dB for the power used for

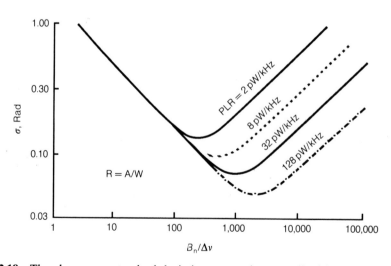

Fig. 2.18 The phase-error standard deviation versus the normalized loop bandwith for several values of the power-to-linewidth ratio.
Source: Kazovsky (1986).

phase locking):

$$\Delta v = 6.2 \times 10^{-4} R_b \qquad (2.59)$$

where R_b is the bit rate.

The other type of phase-locked loop is analyzed in a similar manner. The required linewidth of balanced phase-locked loops for homodyne receivers, in which a part of the transmitter power is used for unmodulated residual carrier transmission, is:

$$\Delta v = 1.2 \times 10^{-5} R_b \qquad (2.60)$$

Balanced phase-locked loops impose more stringent requirements on the laser linewidth than decision-driven loops according to Kazovsky (1986). The performance of the balanced loops is affected by the unmodulated residual carrier, phase noise, shot noise, and the crosstalk between the data detection and phase-lock branches.

2.4 RECEIVER SENSITIVITY COMPARISON

The receiver sensitivity of several coherent detection systems is now compared. To clarify, the signal-to-noise ratio is converted to photons per bit by this equation:

$$E_b = \frac{P_s}{h v R_b} \qquad (2.61)$$

where R_b is the bit rate.

The receiver sensitivities and required spectral linewidth for various coherent optical detection systems are summarized in Table 2.1. The error rates are shown in Fig. 2.19. The received optical power is compared at peak power level. The receiver sensitivity of FSK is improved by 3 dB compared with that of ASK. This is because the signal amplitude of FSK is twice that of ASK while the noise power is also twice. Therefore the signal-to-noise ratio of FSK is twice that of ASK. The receiver sensitivity of DPSK is 3 dB better compared with that of FSK. The bandwidth of DPSK is half that of FSK. Therefore, the sensitivity of DPSK is improved. The receiver sensitivity of optical heterodyne PSK is slightly improved compared with DPSK because of the pure oscillator as the reference signal, which is mentioned in section 2.1.4. The receiver sensitivity difference between heterodyne and homodyne detection is explained by the bandwidth difference.

The required spectral linewidth for envelope detection of ASK and FSK is

Table 2.1 Comparison of optical detection systems

Detection	Modulation	Demodulation	Error rate	Sensitivity (at 10^{-9}) (photons/bit)	Required linewidth	References
Heterodyne	ASK	Envelope	$\frac{1}{2}\exp\left(\frac{\gamma}{4}\right)$	80	$\sim 0.1 R_b$	1
	FSK	Envelope	$\frac{1}{2}\exp\left(\frac{\gamma}{2}\right)$	40	$\sim 0.1 R_b$	1
	CPFSK	Differential detection	$\frac{1}{2}\exp\left[\left(1-\frac{1}{4m}\right)\gamma\right]$	27	$6.8m \times 10^{-3} R_b$	2
	DPSK	Differential detection	$\frac{1}{2}\exp(\gamma)$	20	$3.3 \times 10^{-3} R_b$ $6.3 \times 10^{-3} R_b$ ($\tau_p = 0$)	3
	PSK	Synchronous	$\frac{1}{2}\mathrm{erfc}(\gamma)$	18	$2.43 \times 10^{-3}/\tau_p$ ($\tau_p \gg R_b$) $9.7 \times 10^{-5} R_b$ ($\tau_p = 0$)	4, 5
	QPSK	Synchronous	$\frac{1}{2}\mathrm{erfc}(\gamma)$	18	$2.9 \times 10^{-4}/\tau_p$ ($\tau_p \gg R_b$)	5

	Modulation	Detection	BER	Sensitivity (photons/bit)	Linewidth	Source
Phase diversity	ASK	Envelope	$\frac{1}{2}\exp\left(\frac{\gamma}{4}\right)$	80	$\sim 0.1 R_b$	6
	DPSK	Differential detection	$\frac{1}{2}\exp(\gamma)$	20	$3.3 \times 10^{-3} R_b$; $6.0 \times 10^{-4} R_b$ ($\tau_p = 0$)	6
Homodyne	PSK	Synchronous	$\frac{1}{2}\mathrm{erfc}(\sqrt{2\gamma})$	9	$2.0 \times 10^{-3}/\tau_p$ ($\tau_p \gg R_b$); $9.7 \times 10^{-5} R_b$ ($\tau_p = 0$)	7
	QPSK	Synchronous	$\frac{1}{2}\mathrm{erfc}(\sqrt{\gamma})$	18	$2.9 \times 10^{-4}/\tau_p$ ($\tau_p \gg R_b$)	5

Key:
γ Signal-to-noise ratio
m Modulation index (sensitivity is calculated at m = 1)
R_b Bit rate
τ_p Propagation delay time in PLL

Sources:
1. Garrett and Jacobsen (1986)
2. Iwashita and Matsumoto (1987)
3. Nicholson (1984)
4. Kazovsky (1986)
5. Norimatsu and Iwashita (1992)
6. Davis, et al. (1987)
7. Kazovsky (1985)

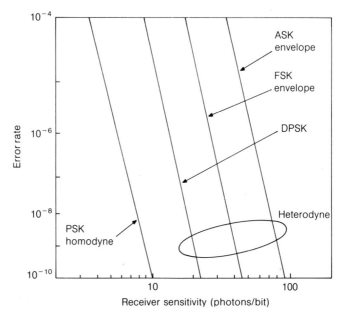

Fig. 2.19 Error rate for various coherent optical detection systems.

determined by the IF filter bandwidth. Namely, the required bandwidth becomes tolerable as the IF bandwidth increases, but the receiver sensitivity decreases. The required linewidths for DPSK and CPFSK depend on the delay time of the differential detectors. The delay time of DPSK systems is 1 bit, so the required linewidth is fixed by the bit rate. On the other hand, as the differential detector delay time of CPFSK systems depends on bit rate and modulation index, the linewidth requirements depend on these two factors. As the synchronous detection linewidth requirements depend on the phase variance margins, they do not depend on the optical heterodyne detection or homodyne detection. However, binary PSK (BPSK) homodyne detection with decision-driven PLL needs some optical power for the PLL. Thus the requirement for heterodyne and homodyne BPSK is different.

The required bandwidth for heterodyne detection except for quadrature PSK (QPSK) is about twice the bit rate (although the required noise bandwidth is the same as the bit rate, in reality twice the bit rate is needed) while that for homodyne detection is about the bit rate. These factors affect the high frequency electrical circuits' construction and the photodiodes' quantum efficiency.

System applications are determined by considering these factors, especially spectral linewidth. The CPFSK system is suitable for long repeater spacing systems because of high sensitivity and the linewidth requirement. The FSK system is suitable for optical frequency division multiplexing (FDM) systems because of its relaxed linewidth requirements and frequency selectivity of optical heterodyne detection. Optical homodyne BPSK detection has the highest receiver sensitivity and required minimum bandwidth at binary systems, but it is difficult to achieve a stable homodyne detection system because of the linewidth requirement.

2.5 POWER SPECTRUM

The power spectral densities are important for evaluating the crosstalk of frequency division multiplexing, the available fiber input power which is restricted by stimulated Brillouin scattering, or high bit rate transmission. The power spectrum of the transmitted signal depends on its modulation format. ASK and PSK signals are the same as a baseband signal; the difference is the carrier component. The ASK signal has a carrier with half the power. The PSK signal, however, has no carrier and the power spectrum is expressed as:

$$S(f) = P_s T \left[\frac{\sin(\pi f T)}{\pi f T} \right]^2 \tag{2.62}$$

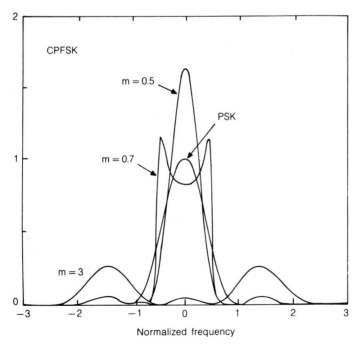

Fig. 2.20 Power/frequency spectrum for PSK and CPFSK transmission for different modulation indexes.

The FSK signal is considered as a continuous phase FSK signal because the FSK signal is mainly produced by laser diode direct current modulation.

The CPFSK power spectrum is expressed as follows.

$$G(\omega) = \frac{2A^2 \sin^2\left\{(\omega - \omega_1)\frac{T}{2}\right\} \sin^2\left\{(\omega - \omega_2)\frac{T}{2}\right\}}{T[1 - 2\cos\{(\omega - \alpha)T\}\cos \beta T + \cos^2 \beta T]} \left(\frac{1}{\omega - \omega_1} - \frac{1}{\omega - \omega_2}\right)^2$$

$$+ \frac{2A^2 \sin^2\left\{(\omega + \omega_1)\frac{T}{2}\right\} \sin^2\left\{(\omega + \omega_2)\frac{T}{2}\right\}}{T[1 - 2\cos\{(\omega + \alpha)T\}\cos \beta T + \cos^2 \beta T]} \left(\frac{1}{\omega + \omega_1} - \frac{1}{\omega + \omega_2}\right)^2 \quad (2.63)$$

where $\alpha = (\omega_2 + \omega_1)/2$ and $\beta = (\omega_2 - \omega_1)/2$.

The power spectrum is shown in Fig. 2.20. The spectrum of CPFSK with a modulation index $m = 0.5$ is narrower than that of PSK. This allows efficient use of frequency bands by frequency division multiplexing. However, the fiber input power is restricted by stimulated Brillouin scattering.

REFERENCES

Blachman, N. M. (1981) The effect of phase error on DPSK error probability, *IEEE Transactions Communications*, **COM-29**, pp. 364–365.

Davis, A. W., Pettitt, M. J., King, J. P. and Wright, S. (1987) Phase diversity techniques for coherent optical receivers, *Journal Lightwave Technology*, **LT-5/4**, pp. 561–572.

Garrett, I. and Jacobsen, G. (1986) Theoretical analysis of heterodyne optical receivers for transmission systems using (semiconductor) lasers with nonnegligible linewidth, *Journal Lightwave Technology*, **LT-4/3**, pp. 323–334.

Glance, B. (1986) Performance of homodyne detection of binary PSK optical signals, *Journal Lightwave Technology*, **LT-4**, pp. 228–235.

Iwashita, K. and Matsumoto, T. (1987) Modulation and detection characteristics of optical contiguous phase FSK transmission systems, *Journal Lightwave Technology*, **LT-5/4**, pp. 452–460.

Kazovsky, L. G. (1985) Decision-driven phase-locked loop for optical homodyne receivers: performance analysis and laser linewidth requirements, *Journal Lightwave Technology*, **LT-3/6**, pp. 1238–1247.

Kazovsky, L. G. (1986) Balanced phase-locked loops for optical homodyne receivers: performance analysis, design considerations, and laser linewidth requirements, *Journal Lightwave Technology*, **LT-4/2**, pp. 182–195, Feb.

Kazovsky, L. G. (1986) Performance analysis and laser linewidth requirements for optical PSK heterodyne communications systems, *Journal Lightwave Technology*, **LT-4/4**, pp. 415–425.

Nicholson, G. (1984) Probability of error for optical heterodyne DPSK systems with quantum phase noise, *Electronics Letters*, **20/24**, pp. 1005–1007.

REFERENCES

Norimatsu, S. and Iwashita, K. (1992) Linewidth requirements for optical synchronous detection systems with nonnegligible loop delay time, *Journal Lightwave Technology*, **10/3**, pp. 341–349.

Okoshi, T., Emura, K., Kikuchi, K. and Kersten, R. T. (1981) Computation of bit-error rate of various heterodyne and coherent-type optical communications schemes, *Journal Optical Communications*, **2/3**, pp. 89–96.

Prabhu, V. K. (1976) PSK performance with imperfect carrier recovery, *IEEE Transactions Aerospace Electronics Systems*, **AES – 12/2**, pp. 275–285.

Schwartz, M., Bennett, W. R. and Stein, S. (1966) *Communication Systems and Techniques*, McGraw-Hill Book Company, New York.

Yamamoto, Y. (1980) Receiver performance evaluation of various digital optical modulation-demodulation systems in the 0.5–10 µm wavelength region, *IEEE Journal Quantum Electronics*, **QE–16/11**, pp. 1251–1259, Nov.

Yamamoto, Y. and Kimura, T. (1981) Coherent optical fiber transmission systems, *IEEE Journal Quantum Electronics*, **QE–17**, pp. 919–934.

3
Coherent transmission technologies

3.1 OPTICAL SOURCE

Katsushi Iwashita

Optical sources for coherent detection utilize lasers. There are several types of laser such as solid state laser, gas laser, and laser diode. Laser diodes are the most promising for practical transmission systems because of their compactness and reliability. However, the main problem is spectral linewidth (determined by phase noise). The spectral linewidth of oscillators for a radio communication system is less than 1 Hz. Therefore, the spectral linewidth of an oscillator is not an important factor. However, Fabry-Perot laser diodes oscillate in several longitudinal modes. The multiple longitudinal mode oscillation can be reduced by distributed feedback (DFB) or distributed Bragg reflector (DBR) laser diode structures. These laser diodes oscillate in only one longitudinal mode but the longitudinal mode has the spectral linewidth resulting from **phase noise**.

The phase noise originates from thermal fluctuation, spontaneous emission fluctuation, and reflective index fluctuation. The fluctuation of carriers originates from carrier number fluctuations.

The spectral linewidth Δv of a laser diode is expressed as follows by the modified Shallow Towns equation in Henry (1982):

$$\Delta v = \frac{hv}{8\pi P}\left(\frac{c^2}{nL_z}\right)\left(\alpha_L L_z + \ln\frac{1}{R}\right)\left(\ln\frac{1}{R}\right)n_{sp}(1+\alpha^2) \tag{3.1}$$

where h is Planck's constant, v is the lasing frequency, n is the reflective index of the laser cavity, P is a facet output power, n_{sp} is the spontaneous emission factor, L_z is the cavity length, R is the facet reflectivity, and α is the spontaneous emission factor.

The measured spectral linewidth dependence on output power is shown in Fig. 3.1. Note that the spectral linewidth narrows with increasing output power. The long cavity is expected to show a narrow spectral linewidth. In fact, however, the spectral linewidth reduction in the high power condition is restricted as shown in Fig. 3.1. The reason for this spectral linewidth limitation is caused by $1/f$

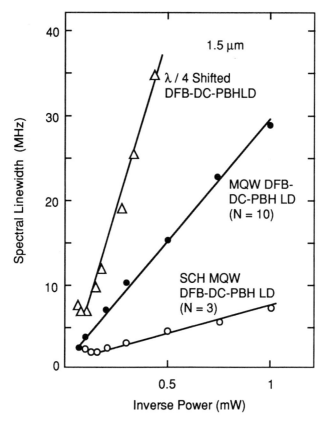

Fig. 3.1 Laser diode spectral linewidth versus optical power. Source: Kobayashi and Mito, 1988.

noise, or by the multi-longitudinal mode operation resulting from the threshold gain difference between the main longitudinal mode and the non-lasing mode (due to spatial hole burning).

The FM noise and the spectral linewidth relation was measured by Kikuchi (1989). The FM noise spectrum consists of white noise and $1/f$ noise. The spectral density of white noise is reduced by increasing the output power, while that of $1/f$ noise is not changed. However, since the $1/f$ noise is a low frequency component, then the influence on coherent detection is analyzed. Above 1 Gbit/s, the influence is not severe.

In a high output power condition, spatial hole burning along the lasing axis is generated, causing the residual carrier to be distributed to the submodes. This results in the spectral linewidth reduction limitation. To avoid the multi-longitudinal operation, several structures were proposed, such as multi electrode DFB

OPTICAL SOURCE

(Yasaka, Hukuda and Ikegami, 1988), multiple phase shift DFB (Kimura and Sugimura, 1987), corrugation pitch DFB (Okai, *et al.*, 1990) and DBR (Matsui, *et al.*, 1991). The corrugation pitch modulated multi-quantum well (MQW) DFB laser structure is shown in Fig. 3.2. By applying a 360 µm phase arranging region in a 1200 µm cavity length, the light intensity along the lasing axis becomes flatter. Then the spectral linewidth of 170 kHz is obtained.

To obtain the required spectral linewidth for coherent lightwave communication, several methods have been proposed, such as an external cavity configuration (Saito, Nilsson and Yamamoto, 1985), and negative frequency feedback (Wyatt and Devlin, 1983). These methods can achieve narrow spectral linewidth sources but it is difficult to obtain long-term stability.

3.2 MODULATORS

Katsushi Iwashita

External modulators are used as amplitude and phase modulators. The electro-optic effect, magneto-optic effect, and acousto-optic effect can be utilized for

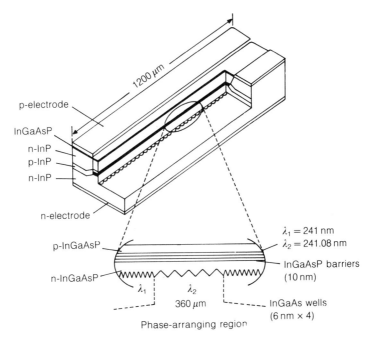

Fig. 3.2 Schematic drawing of narrow linewidth DFB laser.
Source: Okai, *et al.*, 1990.

optical modulators. The electro-optic effect is the most popular for considering modulation response, driving voltage, and optical loss. The materials utilizing this effect are $LiNbO_3$, $LiTaO_3$, and PLZT, KTP etc. These materials have large Pockels effect and Kerr effect. The reflective index can be changed easily by applying an electrical field. The most popular modulator is the $LiNbO_3$ modulator. It utilizes the electro-optic effect to change phase.

3.2.1 Phase modulator

The phase modulator is usually a straight line waveguide modulator with Z-cut Ti-diffused $LiNbO_3$ (Fig. 3.3). The refractive index of the waveguide is changed by the applied voltage, and then the optical phase is modulated. The traveling wave electrode was applied to broaden the modulation bandwidth. The traveling waveguide modulator bandwidth is mainly restricted by the difference between the microwave and optical wave propagation speed.

To avoid this mismatch, several structures are proposed. The configuration of phase modulator to extend the modulation bandwidth using a shielding plate and buffer layer is shown in Fig. 3.4. The waveguide length is 2.7 cm. A phase modulation bandwidth of 12 GHz is achieved. The driving voltage to achieve a 180° phase shift at 10 Gbit/s is 6 V. The insertion loss including input and output fiber is 4 dB. Optical homodyne detection experiments are conducted successfully using this broad bandwidth and low insertion loss modulator as in Norimatsu, Iwashita and Noguchi (1990).

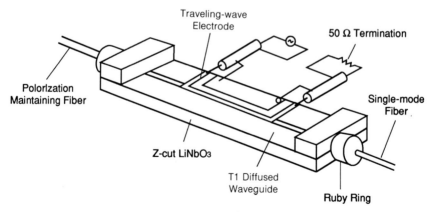

Fig. 3.3 Configuration of an $LiNbO_3$ phase modulator.

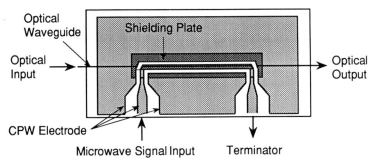

Fig. 3.4 Phase modulator structure.

3.2.2 Direct frequency modulation of laser diodes

The oscillation frequency shift by injection current of a laser diode depends on the modulation frequency. In a low modulation frequency region, less than 1 MHz, the oscillation frequency changes by a thermal effect which is caused by cavity length extension due to increasing the temperature of the active region. As a result, the oscillation frequency shifts to a lower frequency as the current increases. However, in higher frequency regions, the cause of frequency change is the carrier effect which is caused by a change in the carrier-induced refractive index. The oscillation frequency shifts to a higher frequency as the current increases. Therefore, the phase of the modulation response changes between these two effects. This phenomenon is measured by FM response and is shown in Fig. 3.5.

To solve this problem, several structures were proposed such as phase tunable DFB (Yamazaki, *et al.*, 1985), multi-electrode DFB (Yoshikuni and Motosugi, 1986), 2- or 3-section DBR (Murata, Mito and Kobayashi, 1987, and Ishida, Toba and Tohmori, 1989), and lowering dip frequency.

The oscillation frequency of the DFB laser is determined by the facet phase of the grating. Therefore, the optical phase modulator is monolithically integrated in one facet of the DFB laser, and then the grating phase is modulated by controlling the refractive index of the modulator. By using this structure a 300 MHz bandwidth and FM efficiency of 3 GHz/mA is achieved.

The structure of a multi-electrode DFB is shown in Fig. 3.6. The optical power distribution along the DFB laser axis is not constant. The optical power along the cavity is non-uniform, resulting in the lack of carrier density uniformity. By dividing the electrodes along the cavity, the carrier density can be changed, resulting in the optical frequency change. The measured FM response of a multi-electrode DFB laser is shown in Fig. 3.7. Flat and high efficiency response is realized without any phase change.

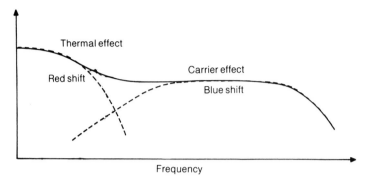

Fig. 3.5 Frequency modulation response of conventional DFB laser.

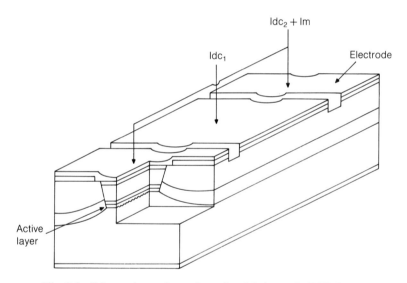

Fig. 3.6 Schematic configuration of multi-electrode DFB laser.

The oscillation frequency of the DBR laser is determined by the reflective index of the DBR region. Through current modulation in this region, the output frequency can be changed. However, since the modulation frequency of this scheme is limited by the carrier life time, it is difficult to realize a high frequency response. In an experiment, a frequency modulation efficiency of 3 GHz/mA has been achieved up to 400 MHz.

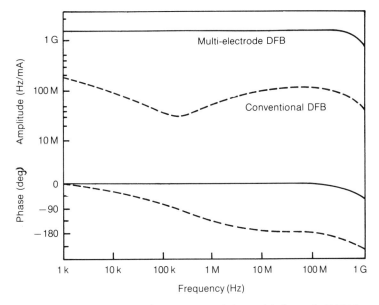

Fig. 3.7 Frequency modulation response of the multi-electrode DFB laser. Source: Nosu and Iwashita, 1988.

3.3 FIBER DISPERSION

Nori Shibata

As bit rate and distance increase in a single-mode fiber transmission system, chromatic dispersion and polarization dispersion become increasingly important. This section describes the properties of chromatic dispersion and polarization dispersion in single-mode fibers.

3.3.1 Chromatic dispersion

Single-mode fibers carry only one propagation mode, though two mutually orthogonal polarization states. If the core is elliptically deformed from the nominally circular cross section, however, the two orthogonal polarization states travel at slightly different group velocities (Kaminow, 1981). Aside from this difficulty, chromatic dispersion in a single-mode fiber is mainly caused by the dispersive properties of the glass material (**material dispersion**), and the dispersion inherent in the guiding process (**waveguide dispersion**) (Maritoson, 1965, and Gloge, 1971). Let us consider a single-mode fiber having a step-index profile.

Chromatic dispersion, D_c, can be approximated as the sum of material dispersion D_m and waveguide dispersion D_w,

$$D_c = \left(\frac{k}{c\lambda}\right)\left(\frac{d^2\beta}{dk^2}\right) \approx D_m + D_w \text{ (ps/km-nm)} \quad (3.2)$$

Here, λ is the wavelength, c the light velocity in free space, $k = 2\pi/\lambda$ the wave number, and β the propagation constant for the guided HE_{11} mode. The material dispersion D_m is expressed as:

$$D_m \approx \left(\frac{k}{c\lambda}\right)\frac{dN_1}{dk} \quad (3.3)$$

where $N_1 = n_1 + k(dn_1/dk)$ is the group index of the core and n_1 is its refractive index. Wavelength dependence of the refractive index $n(\lambda)$ is important to investigate the material dispersion properties, and can be well approximated by a three-term Sellmeier relation (Kabayashi, et al., 1981):

$$n^2(\lambda) - 1 = \sum_{i=1}^{3} \frac{a_i \lambda^2}{\lambda^2 - b_i} \quad (3.4)$$

where a_i and b_i are the coefficients. Wavelength dependence of D_m can be calculated from equation 3.4, and the result for bulk fused silica is shown in Fig. 3.8. The zero material dispersion wavelength λ_{mo} is found to be 1272 nm for silica glass. The wavelength λ_{mo} varies with core dopant materials and dopant concentration.

As an example, material dispersion properties for GeO_2-doped core/silica cladding single-mode fibers are shown in Fig. 3.9. Here the parameter Δ denotes relative index difference defined by:

$$\Delta = \frac{(n_1^2 - n_2^2)}{2n_1^2} \quad (3.5)$$

where n_2 is the refractive index of the cladding. For the silica-based cladding fibers, $\Delta = 0.1\%$ is made by doping approximately 1 mol% GeO_2 into the core region; Δ is proportional to the dopant concentration.

The wavelength λ_{mo} is found to shift to the longer wavelengths as Δ increases. The point of overall vanishing dispersion is shifted by the contribution of waveguide dispersion.

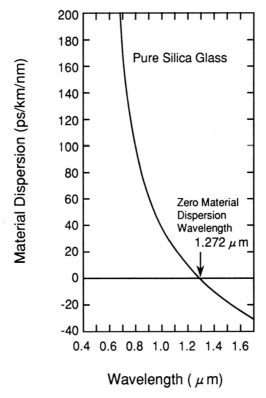

Fig. 3.8 Material dispersion D_m as a function of wavelength for bulk fused silica.

The waveguide dispersion D_W is written in Gloge (1971) as:

$$D_w = \frac{(N_1 - N_2)V}{c\lambda} \frac{d^2(Vb)}{dV^2} \tag{3.6}$$

where

$$V = kn_1 a(2\Delta)^{\frac{1}{2}} \tag{3.7}$$

and

$$b = \frac{\left(\frac{\beta}{k}\right)^2 - n_2^2}{n_1^2 - n_2^2} \tag{3.8}$$

Here, a is the core radius, V the normalized frequency, and $N_2 = n_2 + k(dn_2/dk)$.

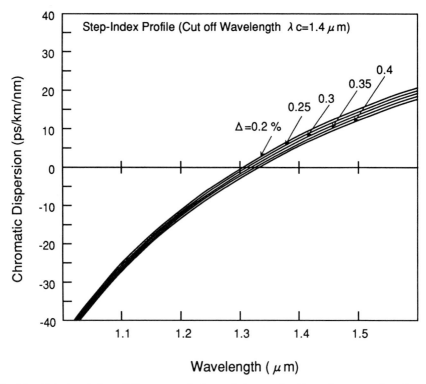

Fig. 3.9 Relative index difference variation of chromatic dispersion as a function of wavelength.

As found from equation 3.6, waveguide dispersion is often characterized by the V-value. For investigating dispersion properties of various types of single-mode fiber with arbitrary index profiles as shown in Fig. 3.10, however, it is convenient to adopt the T-value instead of the V-value. The T-value is defined in Hussey and Pask (1982) as

$$T = 2k^2 \int_0^\infty [n^2(r) - n^2(\infty)] r \, dr \qquad (3.9)$$

The T-value is identical to the V-value for a fiber with a step-index profile. The waveguide dispersion D_w is rewritten using the T-value as follows.

$$D_w = \frac{(N_1 - N_2)T}{c\lambda} \frac{d^2(Tb)}{dT^2} \qquad (3.10)$$

Figure 3.11 illustrates the relationship between normalized waveguide dispersion

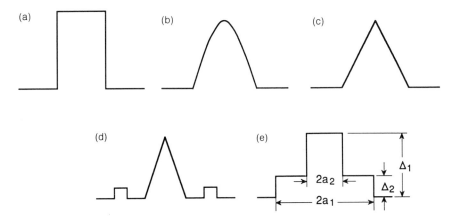

Fig. 3.10 Various types of index profile for single-mode fibers, (a) step, (b) parabolic, (c) triangular, (d) segmented-core and (e) dual-shape core.

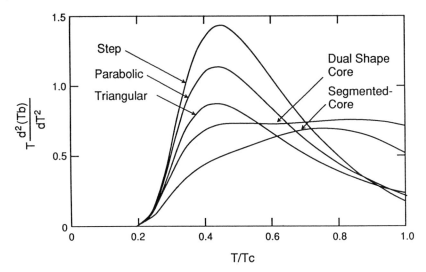

Fig. 3.11 Waveguide dispersion factor $T d^2(Tb)/dT^2$ as a function of the normalized T-value T/T_C. Source: Ohashi, et al., 1987.

$T[d^2(Tb)/dT^2]$ and the T-value normalized by T/T_c, where T_c is the cutoff T-value given at $\lambda = \lambda_c$ (λ_c is the cutoff wavelength) for equation 3.9 as in Ohashi, Kuwaki and Uesugi (1987). A segmented core fiber or dual-shape core fiber can operate as a dispersion-shifted fiber with zero dispersion wavelength in the vicinity of

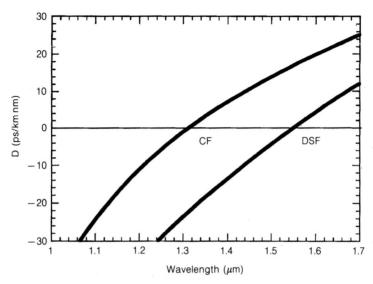

Fig. 3.12 Variation of dispersion parameter D with wavelength for two kinds of fiber. Labels CF, DSF denote conventional and dispersion-shifted fibers, respectively.

1.55 µm, while the conventional single-mode fibers with zero dispersion wavelength around 1.3 µm have step index or parabolic index profiles.

Curve shapes in Fig. 3.11 and excess loss characteristics against bending induced by cabling processes should be taken into account when designing the structure of low-loss dispersion-shifted fibers. Typical dispersion properties of conventional and dispersion-shifted fibers are illustrated in Fig. 3.12. Various techniques have been applied to measure fiber chromatic dispersion in Cohen and Lin (1978), Daikoko and Sugimura (1978), Tateda, Shibata and Seikai (1981) and Cohen (1985).

3.3.2 Polarization dispersion

Polarization dispersion results from modal birefringence in the fiber, and corresponds to the group delay difference between the two orthogonally polarized HE_{11} modes (hereafter called HE_{11}^x and HE_{11}^y modes). The modal birefringence B is generally expressed according to Okamoto, Edahiro and Shibata (1982) as:

$$B = \frac{(\beta_x - \beta_y)}{k} = \frac{(\beta_x^0 - \beta_y^0)}{k} + \frac{(\Gamma_x - \Gamma_y)}{k} \qquad (3.11)$$

where β_x and β_y are the propagation constants for the respective HE_{11}^x and HE_{11}^y

modes, and β_x^0 and β_y^0 are those for the unstressed state. Values of Γ_i for $i = x$ and y are given by:

$$\Gamma_i = \omega\varepsilon_0 \int_0^{2\pi} \int_0^{\infty} \mathbf{E}_i^* \delta\mathbf{K} \mathbf{E}_i r \, dr \, d\theta \quad (3.12)$$

where $\mathbf{E}_i$ and $\delta\mathbf{K}$ represent the electric field vector and the small deviation of the dielectric tensor due to anisotropic thermal stress, respectively, and ω and ε_0 are, respectively, the radian frequency of light and the vacuum value of the permittivity. The first term in equation 3.11 represents geometrical anisotropy B_g, and the second term represents stress-induced birefringence B_s.

Let us consider B_g due to elliptical core deformation. B_g is expressed by Okamoto, Hosaka and Sasaki (1982) as:

$$B_g = n_1 e \Delta^2 G(V) \quad (3.13)$$

where e is the core ellipticity defined by $e = 1 - $ (minor axis a_y/major axis a_x), and $G(V)$ is the normalized phase constant difference. On the other hand, stress-induced birefringence B_s is expressed by Okamoto, Edahiro and Shibata (1982) and Shibata, et al. (1982) as:

$$B_s = C(\sigma_x - \sigma_y)_0 H(V) \quad (3.14)$$

where C is the stress-optical constant, $(\sigma_x - \sigma_y)_0$ the stress difference at the core center, and $H(V)$ the normalized stress difference which is written as:

$$H(V) = \frac{\int_0^{2\pi} \int_0^{\infty} \{\sigma_x(r,\theta) - \sigma_y(r,\theta)\} |E(r,\theta,V)|^2 r \, dr \, d\theta}{\int_0^{2\pi} \int_0^{\infty} (\sigma_x - \sigma_y)_0 |E(r,\theta,V)|^2 r \, dr \, d\theta} \quad (3.15)$$

The stress-induced birefringence B_s is dependent on the V-value as well as the geometrical anisotropy B_g. The polarization dispersion τ_p is then given in Shibata, et al. (1983) by:

$$\tau_p = \left(\frac{1}{c}\right)\frac{d(\beta_x - \beta_y)}{dk} = \left(\frac{1}{c}\right)\left[n_1 e \Delta^2 F(V) + C(\sigma_x - \sigma_y)_0 \left\{M(V) - \left(\frac{\lambda}{C}\right)\left(\frac{dC}{d\lambda}\right)H(V)\right\}\right] \quad (3.16)$$

with

$$F(V) = \frac{d\{VG(V)\}}{dV} \quad (3.17)$$

and

$$M(V) = \frac{d\{VH(V)\}}{dV} \qquad (3.18)$$

Figure 3.13 illustrates $G(V)$, $F(V)$, $H(V)$, and $M(V)$ calculated for elliptical core fibers of $e = 0.2$ and 0.4 as in Okamoto, Hosaka and Sasaki (1982). The validity of equation 3.16 has been experimentally clarified for $e = 0.2$ and 0.4, and the measured values of τ_p at $V = 2$ are approximately 75 ps/km and 120 ps/km, respectively, according to Shibata, *et al.* (1983). The V-value dependence of polarization dispersion can be obtained by measuring τ_p as a function of wavelength. For polarization-preserving fibers with stress-induced anisotropy (Ramaswamy, Kaminow and Kaiser (1978), Katsuyama, Matsumura and Suganuma (1981) and Hosaka, *et al.* (1981)), the wavelength dependence of stress-optical constant C (Sinha, 1978) should be taken into consideration.

In actual single-mode fibers, mode coupling between the two orthogonal polarization states should be taken into account for investigating the length dependence of τ_p. The performance of a very high speed transmission system is limited by the value of propagation delay difference between the two polarization states over the entire transmission distance. When the two polarization modes are excited equally at the input face of the fiber with modal birefringence, the group delay difference $\tau(z)$ between the two polarizations is as expressed in Personik (1971) and Kawakami and Ikeda (1978):

$$\tau(z) = \left(\frac{1}{v_{gx}} - \frac{1}{v_{gy}}\right) \frac{1 - \exp(-2hz)}{2h} \qquad (3.19)$$

where h is the polarization mode coupling coefficient, z the distance along the fiber, and v_{gx} and v_{gy} are the group velocities of the two polarization modes under no polarization mode coupling. At the limit $z \to \infty$, the constant temporal difference is:

$$\tau(\infty) = \frac{\left(\dfrac{1}{v_{gx}} - \dfrac{1}{v_{gy}}\right)}{2h}$$

Transmission limitation due to polarization dispersion is predicted from the propagation delay difference $\tau(\infty)$. As an example, Fig. 3.14 shows the relationship between transmission distance L and the transmission speed B_R for the coherent FSK envelope detection scheme as in Tsubokawa and Sasaki (1988). The limitation was determined by a 0.5 dB degradation in the modulation depth (Sunde 1961). The measured $\tau(\infty)$ was 0.67 ps and the pulse halfwidth per unit fiber-length was determined approximately as $0.053 \text{ ps/m}^{\frac{1}{2}}$. Curve a indicates the limitation due only to polarization dispersion, curves b and c represent the limitation due only

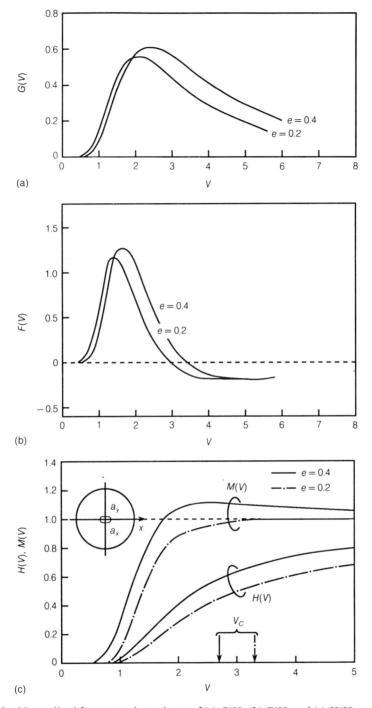

Fig. 3.13 Normalized frequency dependence of (a) $G(V)$, (b) $F(V)$ and (c) $H(V)$ and $M(V)$ for elliptical core fibers with $e = 0.2$ and $e = 0.4$.
Source: Shibata, *et al.*, 1983.

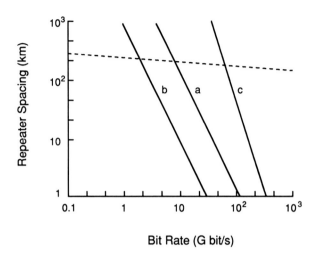

Fig. 3.14 Repeater spacing against data rate for coherent FSK envelope detection system: (a) polarization dispersion limit; (b) chromatic dispersion limit (linear delay distortion 15 ps/km-nm); (c) chromatic dispersion limit (quadratic delay distortion 0.08 ps/km-nm^2). Broken line indicates loss limit for fiber loss of 0.2 dB/km.
Source: Tsubokawa and Sasaki, 1988.

to chromatic dispersion, without polarization dispersion. The broken line shows the fiber loss limit of 0.2 dB/km. The values used to obtain the curves b and c were a chromatic dispersion of 15 ps/km-nm and a dispersion slope of 0.08 ps/km-nm^2, respectively.

Bit error rate and repeater spacing are restricted by polarization dispersion rather than chromatic dispersion, when the coherent system employs dispersion-shifted single-mode fibers (Ainsle, *et al.* (1982), Bhagavatula, Spotz and Love (1984) and Ohashi, *et al.* (1986)) in the 1550 nm wavelength region. Polarization dispersion measurements have been carried out by various techniques as in Shibata (1982), Rashleigh and Ulrich (1978), Monerie, Lamonir and Veunhomme (1980), Shibata, Tateda and Seikai (1982), Shibata, Tsubokawa and Seikai (1984), Bergano, Poole and Wagner (1987), Poole (1989) and Vobian (1990).

3.4 FIBER NONLINEARITIES

Nori Shibata

As the transmitted optical power and number of channels are increased, the various types of nonlinear effect found in silica fiber can dominate system performance under certain conditions. The advent of optical amplifiers raises the

FIBER NONLINEARITIES

problem of fiber nonlinearities, while removing fiber loss as the dominant concern in lightwave communication systems.

Fiber nonlinearities are mainly classified into two categories by Agrawal (1989). One is the stimulated scattering processes such as Raman scattering and Brillouin scattering. The other is the Kerr effect which originates from the third-order nonlinear susceptibility responsible for the phenomena of four-wave mixing, self-phase modulation, and cross-phase modulation. Fiber nonlinearities that are likely to be important in coherent transmission systems include stimulated Raman scattering, stimulated Brillouin scattering, self-phase modulation, cross-phase modulation, and four-wave mixing as in Linke and Gnauck (1988) and Chraplyvy (1990).

3.4.1 Stimulated scattering processes

Stimulated Raman scattering (SRS) and stimulated Brillouin scattering (SBS) are two important nonlinear phenomena in optical fibers. In this section, the features of SRS and SBS are described together because they are quite similar in their origin. The main difference between SRS and SBS is that SRS is the interaction between the lightwave and the vibrational modes of silica molecules, while longitudinal acoustic modes strongly contribute to the interaction with lightwaves in the SBS process. These processes can be described as the scattering of the incident wave (called the pump wave) by molecules and acoustic phonons into a down-shifted wave called the Stokes wave. In the SRS process, energy from the pump wave is transferred to both the co-propagating and counter-propagating Stokes waves as the pump wave travels through the fiber. Stokes waves originate in the optical frequency region downshifted from the pump wave frequency by 13.2 THz and the gain bandwidth is 12 THz. On the other hand, SBS transfers the pump wave energy to a counter-propagating Stokes wave downshifted by 11 GHz and provides narrowband gain with a linewidth of 20 MHz at 1550 nm wavelength. The Brillouin frequency shift and the intrinsic Brillouin bandwidth depend on the system operation wavelength.

(a) Stimulated Raman scattering

In the case of continuous wave operation, the mutual interaction between the pump and Stokes waves is governed by the two coupled equations as follows (Agrawal, 1989).

$$\frac{dI_S}{dz} = g_R I_p I_S - \alpha_S I_S \qquad (3.20)$$

$$\frac{dI_p}{dz} = -\left(\frac{\omega_p}{\omega_S}\right) g_R I_p I_S - \alpha_p I_p \qquad (3.21)$$

where I_S and I_p are the Stokes and pump intensities, respectively, α_S and α_p are fiber attenuation coefficients at the Stokes and pump frequencies of ω_S and ω_p, respectively, and g_R is the Raman gain coefficient. If we assume the solution of the pump wave in the form of $I_p = I_0 \exp(-\alpha_p z)$ as I_0 is the incident pump intensity at $z = 0$, equation 3.20 is rewritten as:

$$\frac{dI_S}{dz} = \{g_R I_0 \exp(-\alpha_p z) - \alpha_S\} I_S \tag{3.22}$$

The solution is easily obtained and the result is:

$$I_S(L) = I_S(0) \exp(g_R I_0 L_{\text{eff}} - \alpha_S L) \tag{3.23}$$

where L is the fiber length, $L_{\text{eff}} = \{1 - \exp(-\alpha_p L)\}/\alpha_p$ is the effective interaction length, and $I_S(0)$ is an input intensity at $z = 0$. The Stokes intensity is found to grow exponentially from spontaneous Raman scattering occurring throughout the fiber length. The Raman threshold is defined as the input power at which the Stokes power becomes equal to the pump power at the fiber output (Smith, 1972). The definition is written as:

$$P_S(L) = P_p(L) = P_0 \exp(-\alpha_p L)$$

where P_0 is the input pump power given by $P_0 = I_0 A_{\text{eff}}$, where A_{eff} is the effective area for the guided HE_{11} mode. Assuming $\alpha_S = \alpha_p = \alpha$, the Raman threshold occurs at a critical pump power given in Smith (1972) as:

$$P_{c-\text{SRS}} = \frac{16 A_{\text{eff}} \gamma}{g_R L_{\text{eff}}} \tag{3.24}$$

where $L_{\text{eff}} = \{1 - \exp(-\alpha L)\}/\alpha$ is the effective length, and γ is the polarization factor related to the polarization directions of the pump and probe waves. For a conventional single-mode fiber, γ takes the value of 2. Under a Gaussian-mode field approximation, the effective area is given by the form of $A_{\text{eff}} = \pi W^2$, where W is the mode field radius depending on the V-value as in Petermann (1983). In a single-channel transmission system, it has been shown that amplification of the Raman scattered light will cause severe degradation when optical power injected into a fiber reaches $P_{c-\text{SRS}}$. The gain peak $g_R \simeq 1 \times 10^{-13}$ m/W at a pump wavelength of 1 μm occurs at a Stokes shift of about 13 THz for silica (Stolen, 1979) and the injected signal power required to produce system degradation is about 1 W according to Chraplyvy (1990) which suggests that SRS will not be a factor in single-channel transmission systems. In WDM systems, the short wavelength channel can act as a pump for longer wavelength channels. The degradation due to SRS is likely to be the most severe for the shortest wavelength channel. Therefore, system performance limits were estimated by calculating the depletion of the shortest wavelength channel by Chraplyvy (1984) and Chraplyvy and Henry (1983). Regarding the effect of SRS in multichannel systems, a power penalty due to Raman-induced crosstalk is described in section 5.3.

(b) Stimulated Brillouin scattering

The impact of SBS is seen when the optical power launched into a fiber is transferred to the backward direction, as is caused by the amplification of spontaneous scattering (Ippen and Stolen (1972), Uesugi, Ikeda and Sasaki (1981) and Cotter (1983)). In such a situation, the forward travelling signal power becomes saturated. Therefore, the important feature of SBS in coherent transmission systems employing narrow line width single-frequency lasers is the level of the critical power injected into the optical fiber. The coupled equations similar to 3.20 and 3.21 are expressed by:

$$\frac{dI_S}{dz} = -g_B I_p I_S + \alpha I_S \qquad (3.25)$$

$$\frac{dI_p}{dz} = -g_B I_p I_S + \alpha I_p \qquad (3.26)$$

where g_B is the Brillouin gain coefficient. The approximated relations of $\omega_S = \omega_p$ and $\alpha_S = \alpha_p = \alpha$ are valid for equations 3.25 and 3.26 because of the relatively small values of Brillouin frequency shift and the fiber loss at the pump frequency is almost identical to that at the Stokes wave frequency.

Comparing equations 3.20 and 3.25, the sign of dI_S/dz is changed to account for the counter-propagating behavior of the Stokes wave with respect to the pump wave. The coupled equation solution leads to exponential growth of the Stokes wave in the backward direction:

$$I_S(0) = I_S(L) \exp(g_B I_0 L_{\text{eff}} - \alpha L) \qquad (3.27)$$

According to a similar procedure with respect to the SRS process, the Brillouin threshold is found to occur at a critical pump power $P_{c-\text{SBS}}$ (Smith, 1972):

$$P_{c-\text{SBS}} = \frac{21 A_{\text{eff}} \gamma}{g_B L_{\text{eff}}} \qquad (3.28)$$

Typical values of $P_{c-\text{SBS}}$ are $< 10\,\text{mW}$ under CW laser operation. The Brillouin frequency shift v_B of the backward scattered wave is given by:

$$v_B = \frac{2\pi n V_a}{\lambda} \qquad (3.29)$$

where V_a is the acoustic velocity. The frequency shift v_B varies inversely with the wavelength λ, and the value is $\sim 11\,\text{GHz}$ at $\lambda = 1.55\,\mu\text{m}$ for silica glass of $V_a = 6000\,\text{m/s}$ and $n = 1.46$. The Brillouin spectral width Δv_B, which is defined as the full width at half maximum, is related to the damping time of acoustic waves. The acoustic waves are assumed to decay exponentially, and the Brillouin gain profile $G_B(v)$ is expressed as a Lorentzian profile given by:

$$G_B(v) = \frac{\left(\dfrac{\Delta v_B}{2}\right)^2 g_B}{(v - v_B)^2 + \left(\dfrac{\Delta v_B}{2}\right)^2} \tag{3.30}$$

The peak value of $G_B(v)$ at $v = v_B$ is given by:

$$G_B(v_B) = g_B = \frac{2\pi n^7 p_{12}^2}{c\lambda^2 \rho V_a \Delta v_B} \tag{3.31}$$

where p_{12} is the Pockels constant and ρ is the glass density. The gain peak value g_B is $4 \times 10^{-11}\,\text{m/W}$. The intrinsic line width Δv_B is proportional to λ^{-2}, and the wavelength dependence of Δv_B has been measured for a bulk glass and a silica-based fiber by Heiman, Hamilton and Hellwarth (1979) and Azuma, et al. (1988).

The Brillouin gain spectra for single-mode fibers are significantly different from those observed for bulk fused silica. The shapes of the spectra vary due to the non-uniformity of core/cladding dopant profiles, fluctuations of refractive index along the fiber length, and the behavior of longitudinal acoustic modes according to Tkach, Chraplyvy and Derosier (1986), Shibata, Waarts and Braun (1987), Jen, et al. (1988) and Shibata, et al. (1988). One of the important features of determining the spectra shape is the behavior of longitudinal acoustic modes guided in the core region (Thomas, et al. (1979), Shibata, Okamoto and Azuma (1989) and Jen, Safaai-Jazi and Farnell (1986)).

Figure 3.15 illustrates Brillouin gain spectra measured at $\lambda = 1550\,\text{nm}$ for fiber A which has a GeO_2-doped core and silica-cladding, and fiber B with a pure silica core and F-doped cladding as in Shibata, et al. (1990). Clearly visible are three Brillouin gain peaks for fiber A, while a strong resonance peak is seen for fiber B. The difference of the two Brillouin gain spectra is due to the guiding condition of the acoustic mode. The longitudinal acoustic modes are guided in the fiber core region when the acoustic velocity for the core material is less than that for the cladding material. Fiber A satisfies the guiding condition, so the three longitudinal acoustic modes guided in the GeO_2-doped core region interact with the HE_{11} mode. The phenomenon for fiber B is similar to that observed for bulk glass.

As seen in Fig. 3.15, the Brillouin spectral shape depends on the core/cladding

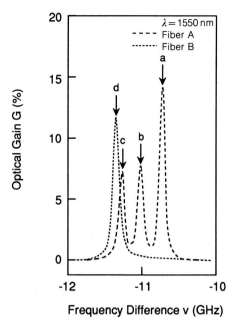

Fig. 3.15 Brillouin gain spectra measured at $\lambda = 1550$ nm for the fiber A with GeO_2-doped core and silica cladding, and fiber B with pure silica core and F-doped cladding. Source: Shibata, et al., 1989.

material composition, and it is easy to predict that the critical power level is also influenced by the fiber type. However, the modulation bandwidth of light sources and the data transmission scheme remain the most important parameters according to Cotter (1983), Aoki, Tajima and Mito (1988), Lichtman, Waarts and Friesem (1989), Bolle, Grosso and Diano (1989), Sugie (1991a) and Sugie (1991b).

The critical power levels for continuous phase frequency shift keying (CPFSK) have been experimentally determined by varying modulation index m and the data rates as in Sugie (1991b). Figure 3.16 shows the input power dependencies of output power at the fiber output end (forward) and the backscattered power at the entrance end (backward). For a carrier wave (CW) input power greater than 5 mW, nonlinear backward and forward SBS transmission characteristics are observed. Forward and backscattered characteristics of 2.488 Gbit/s CPFSK signals with a [1, 0] fixed pattern format for m = 0.73 and pseudo-random pattern format for m = 0.73 and m = 1 are shown in this figure. The backscattered power increases from the Rayleigh scattering level with the CPFSK modulation signal except when m = 0.73. The critical level of the backscattered power is improved by about 7 dB over the value for CW operation by setting m = 1.

Figure 3.17 shows the measured spectrum of a CPFSK signal with m = 1 at 2.488 Gbit/s. The output and the backscattered powers as a function of the input power are shown in Fig. 3.18.

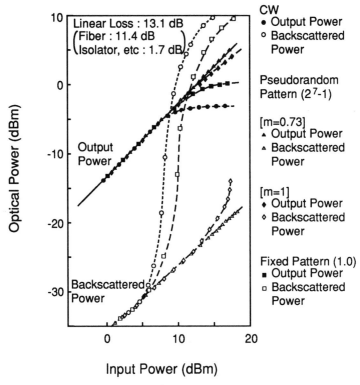

Fig. 3.16 The output power at the fiber output end and the backscattered power at the entrance end versus input power. The optical power for the CW and 2.488 Gbit/s CPFSK signal was measured.
Source: Sugie, 1991b.

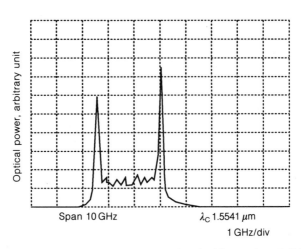

Fig. 3.17 Measured spectrum of CPFSK signal with m = 1 at 2.488 Gbit/s.
Source: Sugie, 1991b.

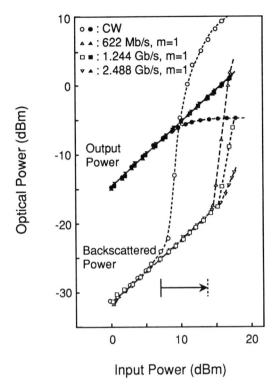

Fig. 3.18 Backscattered power versus input power of the CPFSK signal modulated at bit rates of 622 Mbit/s, 1.244 Gbit/s and 2.488 Gbit/s. The modulation index was m = 1. The output power is also shown.
Source: Sugie, 1991b.

The backscattered power levels were measured under CW operation and CPFSK signals at data rates of 622, 1.244, and 2.488 Gbit/s with m = 1. The critical level increase in the measured backscattered power indicates that SBS is not strongly dependent on data bit rate at m = 1, and the backscattered power at 622, 1.244, and 2.488 Gbit/s starts increasing in the vicinity of input power levels of 13, 13.5 and 14 dBm, respectively. The predicted values of SBS critical power for the respective cases are reported to be 12.5, 12.8, and 12.9 dBm which well describe the measured results.

3.4.2 Optical Kerr effect-based phenomena

Four-wave mixing (FWM), self-phase modulation (SPM), and cross-phase modulation (XPM) are understood by considering the third-order nonlinear polarization

$\mathbf{P}_{NL}$ given by Agrawal (1989) and Shen (1984).

$$\mathbf{P}_{NL} = \varepsilon_0 \chi^{(3)} \vdots \mathbf{EEE} \tag{3.32}$$

where $\mathbf{E}$ is the electric field, $\chi^{(3)}$ is the third-order susceptibility, and ε_0 is the vacuum permittivity. Consider four linearly polarized waves of angular frequencies ω_k ($k = 1, 2, 3$, and 4) injected into an optical fiber and collinearly propagated in the direction z. The electric field in this case can be described as:

$$\mathbf{E} = \mathbf{e}_x \sum_{k=1}^{4} E_k \exp\{-i(\omega_k t + \beta_k z)\} + \text{c.c.} \tag{3.33}$$

where β_k is the propagation constant and $\mathbf{e}_x$ is the unit vector along the x-direction. Substituting equation 3.33 in 3.32, $\mathbf{P}_{NL}$ is assumed to take the form of:

$$\mathbf{P}_{NL} = \mathbf{e}_x \frac{1}{2} \sum_{k=1}^{4} P_k \exp\{-i(\omega_k t + \beta_k z)\} + \text{c.c.} \tag{3.34}$$

The nonlinear polarization P_k of the angular frequency component ω_k can be obtained as the product of three electric fields. For example, P_4 is expressed by Agrawal (1989) as:

$$P_4 = \frac{3\varepsilon_0}{4} \chi^{(3)}_{1111} [\{|E_4|^2 + 2(|E_1|^2 + |E_2|^2 + |E_3|^2)\} E_4$$
$$+ 2E_1 E_2 E_3 \exp(i\Phi_+) + 2E_1 E_2 E_3^* \exp(i\Phi_-) + \cdots] \tag{3.35}$$

where

$$\Phi_+ = (\beta_1 + \beta_2 + \beta_3 - \beta_4)z - (\omega_1 + \omega_2 + \omega_3 - \omega_4)t \tag{3.36}$$

and

$$\Phi_- = (\beta_1 + \beta_2 - \beta_3 - \beta_4)z - (\omega_1 + \omega_2 - \omega_3 - \omega_4)t \tag{3.37}$$

In equation 3.35, $\chi^{(3)}_{1111}$ is one component of the fourth-rank tensor $\chi^{(3)}$. The terms

$$P_{4-1} = \frac{3\varepsilon_0}{4} \chi^{(3)}_{1111} |E_4|^2 E_4$$

and

$$P_{4-2} = \frac{3\varepsilon_0}{2} \{\chi^{(3)}_{1111}(|E_1|^2 + |E_2|^2 + |E_3|^2)\} E_4$$

are responsible for the effects of SPM and XPM, respectively. The remaining terms are responsible for FWM. For the processes of SPM and XPM, special efforts are not necessary to achieve phase matching, while FWM occurs only if the relative phases given by equations 3.36 and 3.37 nearly vanish, and the phase-matching condition must be met with respect to the propagation constant and optical frequency.

The FWM process relevant to the relative phase Φ_+ corresponds to the case in which three waves of ω_1, ω_2, and ω_3 transfer their energy to the wave of ω_4, at the angular frequency $\omega_4 = \omega_1 + \omega_2 + \omega_3$. The phenomenon such as third-harmonic generation corresponds to the case of $\omega_1 = \omega_2 = \omega_3$. However, it is difficult to satisfy the phase-matching conditions for the processes to occur in optical fibers.

The other FWM process with respect to the phase matching occurs at the condition of $\omega_1 + \omega_2 = \omega_3 + \omega_4$, and the relevant phase-matching condition for occurrence is $\Delta\beta = \beta_1 + \beta_2 - \beta_3 - \beta_4 = 0$, which is easily satisfied in the case of $\omega_1 = \omega_2$. This case is very important for multichannel transmission system design. It should be added that the phase-matching condition is automatically satisfied in the case of SRS and SBS as a result of the active participation of the nonlinear medium (Agrawal, 1989). In the following, we describe the processes of FWM, SPM, and XPM for the lightwaves propagating through a single-mode optical fiber.

(a) Four-wave mixing

The four-wave mixing process impacts a coherent transmission system if the system employs frequency division multiplexing (FDM). In optical FDM systems, FWM is the most sensitive to channel spacing and fiber chromatic dispersion, and is likely to impose the most severe restriction on transmitter power. The amount of crosstalk between channels depends on the phase-matching efficiency between the optical waves involved, which in turn depends on frequency spacing, fiber chromatic dispersion, and fiber length as in Shibata, Braun and Waarts (1986 and 1987). The FWM process treated in this section is the partially degenerated case in which $\omega_1 + \omega_2 = \omega_3 + \omega_4$ and $\Delta\beta = \beta_1 + \beta_2 - \beta_3 - \beta_4 = 0$, as described in the previous section.

When three waves of optical frequencies ($j = k$) are injected into a single-mode fiber, nine new frequencies $f_{ijk} = f_i + f_j - f_k$ (subscripts i, j, and k select 1, 2, and 3) are generated through the partially degenerated FWM process.

Figure 3.19 illustrates schematically the three different signal frequencies f_1, f_2, and f_3, and nine new frequencies f_{ijk}. The identification of the origin of the newly generated frequencies is clearly understood from Fig. 3.19. Three signal frequencies and the frequencies generated by FWM overlap if the frequency separations are equal $\Delta f = f_2 - f_1 = f_3 - f_2$, as shown in Fig. 3.19. This leads to the problem of crosstalk in the optical FDM communication systems.

The time-averaged optical power $P_{ijk}(L, \Delta\beta)$ generated through FWM for the frequency component f_{ijk} according to Shibata, Braun and Waarts (1987) and

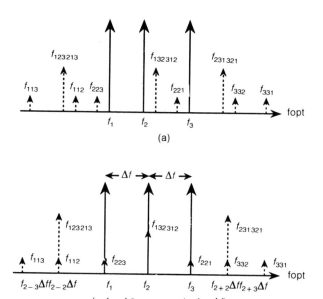

Fig. 3.19 Sketch of three input waves and the nine waves generated through the four-wave mixing process: (a) for different frequency separation; (b) for equal frequency separation with respect to $f_3 - f_2$ and $f_2 - f_1$.

Hill, et al. (1978) is:

$$P_{ijk}(L, \Delta\beta) = \left(\frac{1024\pi^6}{n^4\lambda^2 c^2}\right)(D\chi^{(3)}_{1111})^2\left(\frac{P_i P_j P_k}{A_{\text{eff}}^2}\right)\left|\frac{\exp(i\Delta\beta - \alpha)L - 1}{i\Delta\beta - \alpha}\right|^2 \exp(-\alpha L) \quad (3.38)$$

where L is the fiber length, n the refractive index of the fiber core, λ the wavelength, c the vacuum light velocity, D the degeneracy factor, which can select $D = 1, 3$, and 6, P_i, P_j, and P_k are the input powers injected into a single-mode fiber, and $\Delta\beta(ijk)$ the propagation constant difference which is written as:

$$\Delta\beta(ijk) = \beta_{ijk} + \beta_k - \beta_j - \beta_i \quad (3.39)$$

We expand the propagation constant β in a Taylor series about $\omega_s = 2\pi c/\lambda_s$ and retain terms up to the third order in $\omega - \omega_s$:

$$\beta(\omega) = \beta(\omega_s) + (\omega - \omega_s)\left(\frac{d\beta}{d\omega}\right)_{\omega=\omega_s} + \frac{1}{2}(\omega - \omega_s)^2\left(\frac{d^2\beta}{d\omega^2}\right)_{\omega=\omega_s}$$
$$+ \frac{1}{6}(\omega - \omega_s)^3\left(\frac{d^3\beta}{d\omega^3}\right)_{\omega=\omega_s}$$

$$= \omega_s/v_p + (\omega - \omega_s)/v_g + (\omega - \omega_s)^2 \left(\frac{\lambda_s^2 D_c}{4\pi c}\right)$$
$$+ (\omega - \omega_s)^3 \left(\frac{\lambda_s^4}{24\pi^2 c^2}\right) \left\{\left(\frac{2D_c}{\lambda_s}\right) + \left(\frac{dD_c}{d\lambda}\right)\right\} \quad (3.40)$$

where v_p and v_g are the phase and group velocities of the HE_{11} mode beam traversing the fiber, respectively, and D_c is the fiber chromatic dispersion given by:

$$D_c = -\left(\frac{\omega_s^2}{2\pi c}\right)\left(\frac{d^2\beta}{d\omega^2}\right)_{\omega = \omega_s} \quad (3.41)$$

In equation 3.40, D_c dominates, and the contribution of the dispersion slope $dD_c/d\lambda$ can be neglected at wavelengths far from the zero chromatic dispersion wavelengths λ_0 of around 1.3 and 1.55 μm. At the zero chromatic dispersion wavelength, $D_c = 0$ and the slope $dD_c/d\lambda$ must be included. For two cases, where the operation wavelength is far from λ_0 and close to λ_0, $\Delta\beta(ijk)$ is expressed respectively by Shibata, Braun and Waarts (1987), Hill, et al. (1978) and Inoue and Toba (1992) as:

$$\Delta\beta(ijk) = \left(\frac{2\pi\lambda_s D_c}{c}\right)\Delta f_{ik}\Delta f_{jk} \quad (3.42)$$

$$\Delta\beta(ijk) = \left(\frac{\pi\lambda_s^4}{c^2}\right)\left(\frac{dD_c}{d\lambda}\right)\Delta f_{ik}\Delta f_{jk}(f_i + f_j - 2f_0) \quad (3.43)$$

where $\Delta f_{mn} = f_m - f_n$ ($m, n = i, j, k$) and $f_0 = c/\lambda_0$. The wave generation efficiency η with respect to phase mismatch $\Delta\beta(ijk)L$ can be expressed as in Shibata, Braun and Waarts (1987):

$$\eta \equiv \frac{P_{ijk}(L, \Delta\beta)}{P_{ijk}(L, \Delta\beta = 0)} = \left\{\frac{\alpha^2}{\alpha^2 + \Delta\beta(ijk)^2}\right\}\left[1 + \frac{4\exp(-\alpha L)\sin^2\{\Delta\beta(ijk)L/2\}}{\{1 - \exp(-\alpha L)\}^2}\right] \quad (3.43a)$$

From equations 3.42 and 3.43 the efficiency η is described as the equivalent frequency separation defined as $\Delta f_{eq}(ijk) = (\Delta f_{ik}\Delta f_{jk})^{\frac{1}{2}}$ which is related to the channel spacing in multichannel transmission systems. It is noted that the degeneracy factor $D = 1, 3,$ or 6 depends on whether three, two, or none of the frequencies $f_1, f_2,$ and f_3 are the same. For the respective cases of $f_i = f_j \neq f_k$ and $f_i \neq f_j \neq f_k$, $P_{ijk}(L, \Delta\beta)$ is then rewritten as:

$$P_{ijk}(L, \Delta\beta) = \eta\left(\frac{1024\pi^6}{n^4\lambda^2 c^2}\right)(3\chi_{1111}^{(3)})^2\left(\frac{L_{eff}}{A_{eff}}\right)^2 P_i^2 P_k \exp(-\alpha L) \quad (3.44)$$

and

$$P_{ijk}(L, \Delta\beta) = \eta \left(\frac{1024\pi^6}{n^4\lambda^2 c^2}\right)(6\chi^{(3)}_{1111})^2 \left(\frac{L_{eff}}{A_{eff}}\right)^2 P_i P_j P_k \exp(-\alpha L) \quad (3.45)$$

In general, transmission systems are designed at the zero chromatic dispersion wavelengths λ_0 of 1.3 and 1.55 µm. The FWM efficiency behavior at $\lambda = \lambda_0$ is very important in investigating the FWM-induced crosstalk problem. The phase matching condition is found from equation 3.43 to be automatically satisfied if λ_0 is located at the middle of f_i and f_j, that is, $f_i + f_j - 2f_0 = 0$.

Optical frequency allocations that satisfy the phase-matching condition in the zero chromatic dispersion wavelength region are shown in Fig. 3.20(a) and (b) which correspond to a partially degenerate case and a completely nondegenerate case, respectively. The dispersion slope can be approximated as a constant value of about 0.07 ps/km-nm² in the wavelength range of $\lambda_0 + 10$ nm. For the frequency allocations shown in Fig. 3.20, the FWM efficiency of $\eta = 100\%$ is valid over the linear approximation wavelength region.

From the viewpoint of multichannel transmission system design, the channel frequency allocation in the vicinity of λ_0 should be avoided because of the increase of FWM-induced crosstalk, while the frequency allocation is very suitable for useful applications of fiber FWM such as:

all-optical demultiplexing (Anderkson, et al., 1991);
picosecond optical sampling (Anderkson, 1991);
frequency convertor (Inoue and Toba, 1992).

For the frequency allocation, except for the case shown in Fig. 3.20, η is expressed (Shibata, Braun and Waarts 1986) as a function of $\Delta f_{eq}(ijk)$.

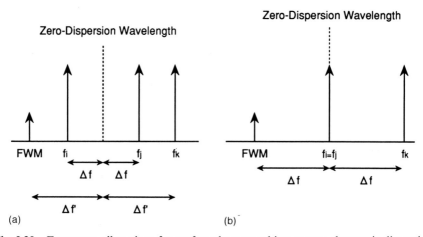

Fig. 3.20 Frequency allocations for perfect phase matching at zero chromatic dispersion wavelengths.

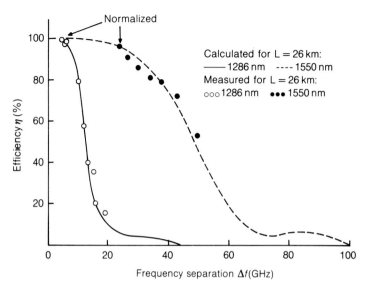

Fig. 3.21 Channel separation dependence of wave generation efficiency through four-wave mixing at $\lambda = 1286$ nm and 1550 nm, for 26 km-long dispersion-shifted single-mode fiber.
Source: Shibata, et al., 1988.

Figure 3.21 shows the efficiency measured at 1265 and 1550 nm wavelengths for a 26 km-long dispersion-shifted single-mode fiber with $\lambda_0 = 1565$ nm (Shibata, et al., 1988). In this case, two CW waves were injected into the fiber such that $\Delta f_{eq}(ijk) = \Delta f = f_2 - f_1$. The FWM powers of the frequency components f_{112} and f_{221} were measured as a function of the frequency separation Δf which was varied from 3.8 to 19 GHz, and from 22 to 50 GHz, with $P_1 = 0.28$ and 0.28 mW, and $P_2 = 0.56$ and 0.28 mW at 1265 and 1550 nm, respectively. The generated powers of P_{112} and P_{221} decrease with the increase in Δf because of an increase in phase mismatch. The experimental results reflect the theoretically obtained curves.

The efficiency η behavior was first verified at the 0.8 μm wavelength region, and the variation of optical power $P_{ijk}(L, \Delta \beta)$ as a function of the input power levels was also investigated by Shibata, Braun and Waarts (1987). Three CW waves with frequency separations of $f_2 - f_1 = 17.2$ GHz and $f_3 - f_2 = 11.0$ GHz were propagated through a single-mode fiber with a length of 3.5 km. The generated powers of the frequency components f_{332} and f_{231} were measured by varying the input power P_3.

Figure 3.22 shows the measured variation for $P_1 = 0.43$ mW and $P_2 = 0.14$ W. The generated power P_{231} varies linearly with P_3 but quadratically for the power P_{332} according to Shibata, Braun and Waarts (1987).

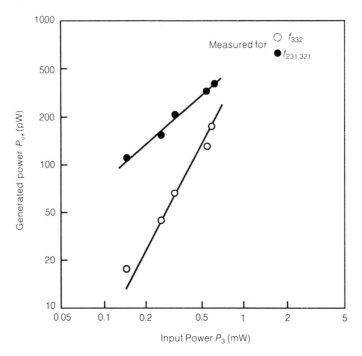

Fig. 3.22 Measured output powers with frequency components of f_{332} and $f_{231,321}$ as a function of P_3.
Source: Shibata, et al., 1987.

(b) Self-phase modulation and cross phase modulation

SPM and XPM are the nonlinear phenomena that occur as a result of the intensity dependence of the refractive index in Agrawal (1989) and Chraplyvy (1990). SPM refers to the self-induced phase shift experienced by an optical field propagation through an optical fiber and XPM refers to the nonlinear phase shift of a light field induced by a co-propagating field at a different wavelength (Agrawal, 1989).

The main effect of SPM is known to be responsible for the spectral broadening of optical pulses propagating through the optical fiber. The important feature of SPM in the single-channel transmission system is that SPM converts optical power fluctuations in the lightwave into phase fluctuations in the same wave. In coherent transmission systems employing a phase shift keyed (PSK) scheme, the effects of SPM and fiber dispersion act together to influence the system performance of long-haul transmission systems employing optical amplifiers at high bit rates. On the other hand, XPM converts the optical power fluctuations in an optical channel to phase fluctuations in other channels.

The XPM effects have been observed in a two-channel experiment using laser diodes operated at 1.3 and 1.5 μm wavelengths by Chraplyvy and Stone (1984), and the influence on a two-channel PSK homodyne transmission system has been experimentally studied by Norimatsu and Iwashita (1991a).

The intensity dependence of the refractive index resulting from the contribution of $\chi^{(3)}$ is expressed as:

$$n = n_0 + n_2|E|^2 \qquad (3.46)$$

where $|E|^2$ is the optical intensity injected into the fiber, and n_2 is the nonlinear index coefficient given by:

$$n_2 = \frac{3}{8n}\chi^{(3)}_{1111} \qquad (3.47)$$

Equation 3.47 assumes the electric field to be linearly polarized. The phase of light after propagating through an optical fiber of length L is written in Chraplyvy (1990) as:

$$\phi(L) = \frac{2\pi}{\lambda}n_0 L + \frac{2\pi}{\lambda}n_2|E|^2 L_{\text{eff}} \qquad (3.48)$$

The intensity-dependent nonlinear phase shifts for the optical field of E_1 are expressed as:

$$\phi_{\text{SPM}} = \frac{2\pi}{\lambda}n_2|E_1|^2 L_{\text{eff}} \qquad (3.49)$$

and

$$\phi_{\text{XPM}} = \frac{2\pi}{\lambda}n_2(|E_1|^2 + 2|E_2|^2)L_{\text{eff}} \qquad (3.50)$$

for SPM and XPM, respectively. In the case of XPM, two optical fields E_1 and E_2 with different frequencies are injected into the optical fiber, and they co-propagate through the fiber. As seen from equations 3.49 and 3.50, optical power fluctuation produces changes in the phase, and long-haul transmission systems employing PSK schemes will be influenced by the SPM and XPM. In single-channel systems, the phase change in the detected signal caused by the nonlinear refractive index is given by Chraplyvy (1990) as:

$$\sigma_\phi = 0.035\sigma_p \qquad (3.51)$$

where σ_ϕ is the rms phase fluctuation in radians and σ_p is the rms power fluctuation in milliwatts. XPM effects due to power fluctuations in other channels are seen in multichannel transmission systems, in addition to SPM. In an N channel system, the rms phase fluctuations in a certain channel due to the power fluctuation in the other channel are, according to Chraplyvy (1990):

$$\sigma_\phi = 0.07\sqrt{N}\sigma_p \tag{3.52}$$

The XPM contribution can become large as the number of channels increases, and the SPM contribution is negligibly small.

3.5 HETERODYNE RECEIVER

Katsushi Iwashita

The receiver sensitivity of a coherent detection system is mainly determined by the following factors:

1. modulation/demodulation scheme;
2. optical source spectral linewidth;
3. photodiode quantum efficiency;
4. modulator response;
5. thermal noise;
6. optical source excess noise;
7. receiver bandwidth;
8. demodulation process (electrical mixer flatness);
9. intersymbol interference;
10. others.

Factors 1 and 2 were discussed in section 3.2. Here, we will discuss thermal noise (5) and optical source excess noise (6).

3.5.1 Thermal noise

The receiver sensitivity is restricted by insufficient local oscillator power and thermal noise. The receiver sensitivity degradation due to restricted local power is expressed as follows.

$$P_D = \frac{N_{sh}}{N_{sh} + N_{th}} \tag{3.53}$$

where N_{sh} is the shot noise power and N_{th} is the thermal noise power.

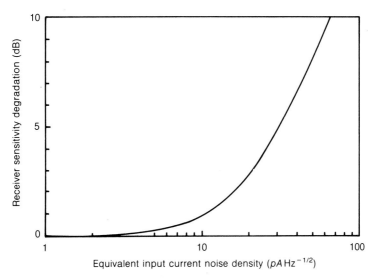

Fig. 3.23 Receiver sensitivity degradation due to thermal noise, where local oscillator optical power is 0 dB and quantum efficiency is 100%. Thermal noise is assumed to be white noise.

The influence of thermal noise on receiver sensitivity is shown in Fig. 3.23. By increasing the local oscillator power, the thermal noise is ignored and shot noise limit detection can be realized. The thermal noise of the receiver consists of $1/f$ noise, flat noise, and frequency squared noise.

The thermal noise density using FET receivers is expressed as follows in Personik (1973):

$$\langle i^2 \rangle = \frac{4kT}{R_{in}} + 2eI_g + \frac{4kT\Gamma}{g_m}(2\pi C_T)^2 f^2 \quad (3.54)$$

where k is Boltzmann's constant, T is the absolute temperature, g_m is the transconductance, R_{in} is the input resistance, e is the electron charge, I_g is the gate current, C_T is the input capacitance and Γ is the numerical noise parameter (1.75 for GaAs MESFETs). Here, we ignored $1/f$ noise because, in heterodyne detection, the intermediate frequency is utilized.

The first and second terms are constant with frequency; however, the third term has f^2 dependence. Therefore, $1/f^2$ noise can be the main thermal noise factor in high frequency receivers. To reduce f^2 noise, several methods are proposed, such as capacitive peaking (Iwashita and Matsumoto (1987)) and inductor peaking MMIC configuration as in Ohkawa (1988).

3.5.2 Balanced receiver

Next we will describe the relative intensity noise (RIN) cancellation using a balanced receiver. The relative intensity noise of a laser is expressed as follows.

$$\text{RIN} = \frac{\langle \delta P(\omega)^2 \rangle}{P} \quad (3.55)$$

where P is the optical source power, and $\delta P(\omega)$ is the power fluctuation. RIN influence on receiver sensitivity is expressed as follows.

$$S/N = (S/N)_{\text{ideal}} \frac{N_{\text{sh}}}{N_{\text{sh}} + N_{\text{th}} + \text{RIN} \cdot \left\{ \frac{\eta e}{h\nu} (P_L + P_S) \right\}^2} \quad (3.56)$$

where $(S/N)_{\text{ideal}}$ is the ideal signal-to-noise ratio.

If we assume RIN = $-160\,\text{dB/Hz}$ at $6\,\text{pA}/\sqrt{\text{Hz}}$ as shown in Fig. 3.24, then more than 0 dBm local oscillator power affects the receiver sensitivity.

This relative intensity noise can be cancelled with a balanced receiver. The balanced receiver configuration is shown in Fig. 3.25. The transmitted signal is combined with local oscillator light by adjusting the polarization. The two

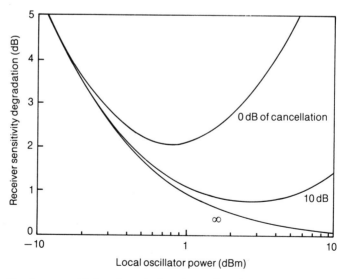

Fig. 3.24 Receiver sensitivity degradation as a function of excess intensity noise cancellation.

HETERODYNE RECEIVER

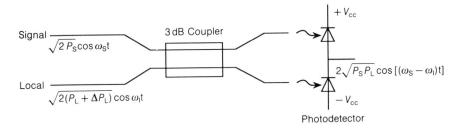

Fig. 3.25 Balanced receiver configuration.

coupled output signals are expressed as follows.

$$E_1(t) = \frac{P_S}{2} + \frac{P_L}{2} + \sqrt{P_S P_L}\{\cos((\omega_S - \omega_L)t + \phi_S - \phi_L + \varphi(t))\} \quad (3.57)$$

$$E_2(t) = \frac{P_S}{2} + \frac{P_L}{2} - \sqrt{P_S P_L}\{\cos((\omega_S - \omega_L)t + \phi_S - \phi_L + \varphi(t))\} \quad (3.58)$$

The two coupled signals are detected by photodetectors. The first and second terms produce shot noise and excess noise of optical sources. The shot noise resulting from the two photodiodes is an independent process. Therefore, the two noise powers are added to each other. However, the excess noise of the optical sources is the same noise; thus the two noise powers are cancelled by subtracting each other. The third term is the signal but the signs of the two signals are opposite; therefore, the two signals are added and the signal voltage doubles. Therefore, the signal amplitude is doubled and the shot noise power is doubled. The signal-to-noise ratio is improved by 3 dB compared with single detection. Moreover, the local oscillator light has an intensity excess noise, and can be cancelled by the subtraction.

Heterodyne detection systems require a wideband and a high frequency optical receiver compared with a direct detection receiver because heterodyne systems require a bandwidth three times the bit rate. To achieve shot noise limit detection, the receiver configuration is complicated because it needs a coupler and two receivers. Moreover, it requires precise control between the two paths at a receiver to achieve good cancellation. Therefore, it is very useful to realize monolithic integration within the receiver such as twin PIN photodiodes, and a matching integrated circuit.

Figure 3.26 shows the optical receiver using twin PIN photodiodes and a monolithic microwave IC with solder bonding as in Takachio, et al. (1990). An InGaAs twin PIN photodiode is used as the photodetector. The twin photodiode can suppress the local oscillator intensity noise by connecting a balanced configuration. Moreover, it makes efficient use of low power and operates close to the shot noise limit. The distributed amplifier was fabricated using GaAs

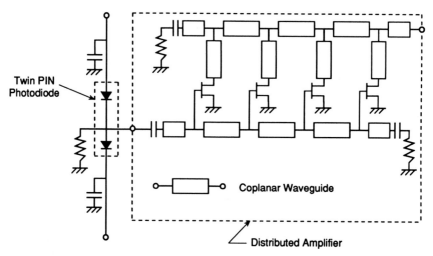

Fig. 3.26 Optical receiver using twin PIN photodiode and monolithic IC.
Source: Takachio, et al., 1990.

MESFETs. Even though distributed amplifiers have a wide frequency bandwidth, the optical receiver bandwidth is restricted by the stray inductances and capacitances that accompany the connection between a photodiode and the amplifier. The twin PIN photodiode and the distributed amplifier are connected by a solder bonding technique to reduce the parasitic effects.

The frequency response and noise characteristics of the optical receiver are shown in Fig. 3.27. The measured minimum input noise current was $19.4 \, \text{pA}/\sqrt{\text{Hz}}$ at 2 GHz.

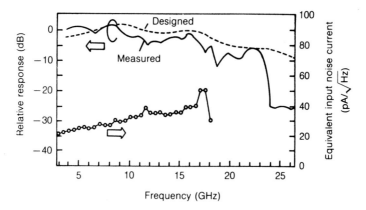

Fig. 3.27 Frequency response and noise characteristics of the optical receiver.
Source: Takachio, et al., 1990.

The twin PIN photodiode is a promising way to get good cancellation because it is easy to obtain similar characteristic photodiodes.

3.5.3 Phase diversity

The IF frequency in heterodyne detection is determined by considering the following factors:

- frequency response of photodiode;
- frequency response of electrical circuits (amplifier, mixer);
- thermal noise;
- double frequency component at demodulator output interference;
- demodulation scheme restriction (such as DPSK and CPFSK).

To lower the IF frequency, phase diversity detection which avoids the need for phase locking is proposed by Davis, *et al.* (1987). The advantage of baseband processing can be retained by employing a system which gives phase diversity output from two or more arms.

The configuration of three-port phase diversity is shown in Fig. 3.28. The three-port signals are as follows:

$$V_2(t) = \frac{\eta e}{h\nu} \left[\frac{P_S}{3} + \frac{P_L}{3} + \frac{2}{3}\sqrt{P_S P_L} \left\{ \cos\left((\omega_S - \omega_L)t + \phi_S - \phi_L + \varphi(t) + \frac{2n\pi}{3} \right) \right\} \right]$$

(3.59)

for $n = 0, 1, 2$.

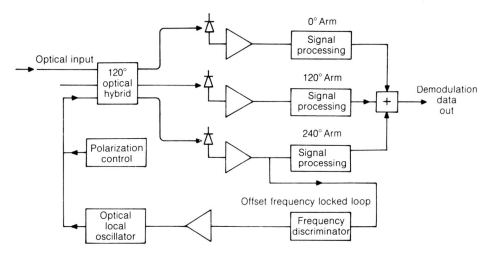

Fig. 3.28 Three-port phase diversity receiver configuration.

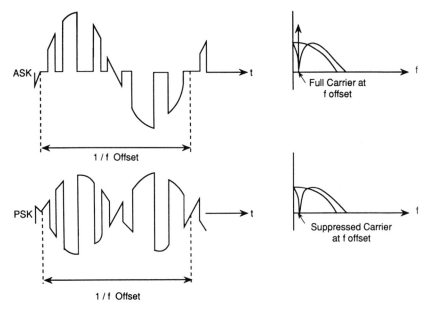

Fig. 3.29 Output waveforms from phase diversity receiver.

The output is shown in Fig. 3.29. In phase diversity detection, the intermediate frequency is set lower than the bit rate, namely below $\frac{1}{10}$ the bit rate.

The signal in each arm can be processed to recover ASK, FSK, and DPSK modulation. Each arm signal becomes zero, but when one is zero, the others are still present. Summing the outputs gives a full demodulated signal.

Because the signal and local oscillator power are split into two or three arms, the receiver sensitivity is 3 dB (2 times) or 4.8 dB (3 times) worse compared with homodyne detection, respectively. Therefore, the receiver sensitivity for two-phase detection is the same as for heterodyne detection. The receiver sensitivity for three-phase processing is worse by 1.8 dB.

The optical hybrid that combines local oscillator power and signal power is a key component in a phase diversity system as shown in Fig. 3.30. A six-port fiber coupler is made by fusing three fibers until equal power appears at each output port. Then a 120° phase relation is obtained. This coupler does not require phase adjustment, which is useful. An eight-port coupler was realized by 4 couplers with control in the path length in the cross coupler. This type was not stable and required automatic phase control. An alternative form consists of a beam splitter and two polarization beam splitters as shown in Fig. 3.30(c). To obtain this phase relation, one of the two signals (the received signal or the local oscillator signal) must be linearly polarized while the other must be circularly polarized.

To describe the operation of the optical hybrid, let us assume the local oscillator

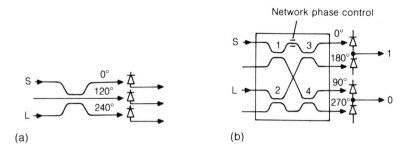

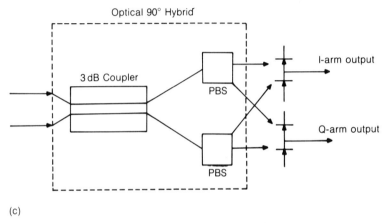

Fig. 3.30 Optical hybrid configuration: (a) six-port fiber coupler-type optical hybrid; (b) eight-port optical hybrid with 4 couplers; (c) eight-port optical hybrid with a half mirror and polarization beam splitters.

signal is circularly polarized:

$$E_L(t) = (\sqrt{P_L}\cos\omega_L t)_x + (\sqrt{P_L}\sin\omega_L t)_y \tag{3.60}$$

where x denotes the polarization component and y denotes the orthogonal polarization component.

Let us assume that the signal is linearly polarized; then the signal can be written as:

$$E_S(t) = (\sqrt{P_S}\cos\omega_S t)_x + (\sqrt{P_S}\cos\omega_S t)_y \tag{3.61}$$

The I-arm signal which derives from the x-polarization component is given as:

$$I_1(t) = \frac{\eta e}{2h\nu}\sqrt{P_S P_L}\cos(\omega_L - \omega_S)t \tag{3.62}$$

In the same way, the Q-arm signal is derived as:

$$I_Q(t) = \frac{\eta e}{2h\nu}\sqrt{P_S P_L}\sin(\omega_L - \omega_S)t \tag{3.63}$$

Then the 90° different signal is obtained. In fact, there are two outputs from the beam splitter. The other outputs are out-of-phase; thus these outputs are utilized as a balanced receiver as shown in Fig. 3.30(c).

3.5.4 Image rejection mixer

Heterodyne detection is used to downconvert a high frequency to a lower intermediate frequency where signal processing is easier. A conventional heterodyne receiver detects the signal received at $f_L + f_{IF}$ or at $f_L - f_{IF}$. One of these signals is unwanted and defined as the image frequency. The presence of the image signal does not affect the conventional receiver performance as long as the system noise is generated in the receiver. However, image frequency cannot be ignored in the coherent transmission system using a densely spaced frequency division multiplexing (FDM) system or an in-line repeatered optical amplifier system.

An image rejection mixer was proposed by Glance (1986). The image rejection mixer detects a desired signal such as a higher frequency than that of the local oscillator while rejecting lower frequencies (image). It can distinguish between these two signals. Therefore, it is useful for frequency division multiplexing with narrow spaced frequency signals. The configuration is shown in Fig. 3.31. This utilizes optical and electrical 90° hybrids.

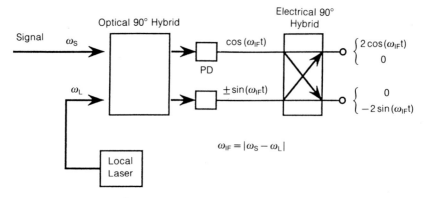

Fig. 3.31 Image rejection mixture configuration.

The output of an optical 90° hybrid is expressed as follows:

$$I_I(t) = \frac{\eta e}{2h\nu}\sqrt{P_S P_L}\cos(\omega_L - \omega_S)t \qquad (3.64)$$

$$I_Q(t) = \frac{\eta e}{2h\nu}\sqrt{P_S P_L}\sin(\omega_L - \omega_S)t \qquad (3.65)$$

The output signal is out-of-phase. If we combine these signals using an electrical 90° hybrid, the output signal on one output port is added in-phase but the other port is out-of-phase. Then the output of the port with out-of-phase signal is cancelled. These conditions will depend on the signal and local oscillator frequency relationship.

3.6 TRANSMISSION CHARACTERISTICS AND DELAY EQUALIZATION

Katsushi Iwashita

Transmission characteristics depend upon the following factors:

- fiber loss;
- chromatic dispersion;
- fiber nonlinearity;
- polarization dispersion.

In this section, we will estimate the degradation due to fiber chromatic dispersion.

3.6.1 Transmission characteristics

The transmission distance limitation due to fiber chromatic dispersion was estimated by Elrafaie, *et al.* (1988) and Iwashita and Takachio (1989). The results are shown in Fig. 3.32. The transmission distance limitation due to linear chromatic dispersion is proportional to the square of the bit rate and dispersion. The transmission distance depends on the modulation/demodulation format.

3.6.2 Delay equalization

Chromatic dispersion effects can be reduced in the electrical domain for direct detection systems, but nonlinear techniques are required. On the other hand, in coherent detection systems, the distortion can be easily compensated in the

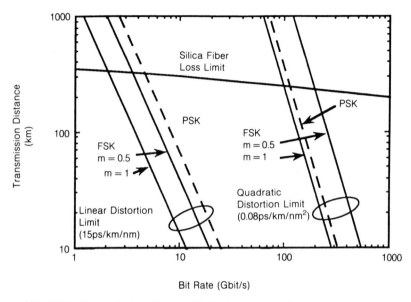

Fig. 3.32 Transmission distance limitation due to chromatic dispersion.

electrical domain because the output signal has a linear relation to the electrical field of the optical signal. Electrical dispersion compensation techniques have been developed for radio communication systems that have the advantages of lack of sensitivity degradation.

When a 1.3 μm zero-dispersion fiber is used at the wavelength of 1.55 μm, high frequency components propagate faster than low frequency ones, as shown in Fig. 3.33. If the local oscillator frequency is set lower than that of the signal, as

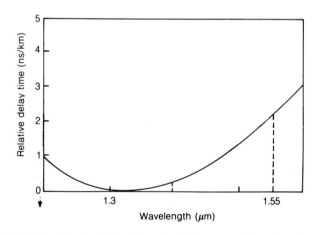

Fig. 3.33 Relative delay time of a 1.3 μm zero dispersion fiber.

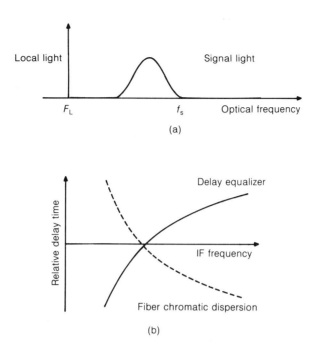

Fig. 3.34 Schematic configuration of chromatic dispersion compensation in heterodyne detection. Frequency arrangement of signal and local detection: (a) frequency arrangement of signal and local oscillator laser; (b) relative delay time of fiber and delay equalizer.

shown Fig. 3.34(a), the relative delay time in the IF band becomes longer as the IF frequency increases because the fiber delay time is directly converted to IF delay time. The chromatic dispersion can be compensated by using a delay equalizer in the IF band which has a delay characteristic opposite to the fiber chromatic dispersion, as shown in Fig. 3.34(b).

If the local oscillator frequency is set higher than that of the signal, the relative delay time in the IF band decreases as the IF frequency increases because the frequency relation is reversed by heterodyne detection. The chromatic dispersion can be compensated by using a delay equalizer in the IF band which has the same delay characteristic as the fiber chromatic dispersion.

Delay equalizers such as waveguide equalizers or coupled equalizers have already been developed for millimeter-wave communication systems. Chromatic dispersion of optical fiber is lower than that of millimeter-wave waveguides. Therefore, fiber chromatic dispersion is important at more than several gigabits per second. For example, as waveguide dispersion is around 2 ns/GHz/km, the relative delay time is 20 ns/GHz with 10 km transmission. Dispersion is significant at around 400 Mbit/s. The bandwidth of developed delay equalizers for millimeter-wave communication systems is approximately 1 GHz. On the other hand, the dispersion of fiber is around 25 ps/GHz with 200 km single mode fibers. Since the influence of dispersion appears at data rates of more than 5 Gbit/s, the bandwidth of delay equalizers must be more than 10 GHz.

The other requirements for delay equalizers are small size, low loss, and matched impedance with other devices. Three types of delay equalizers for optical communications systems have been proposed: microstrip line delay equalizer (Takachio and Iwashita, 1988), all-pass filter type delay equalizer (Priest and Giallorenzi, 1987), and waveguide equalizer (Winter, 1989).

The microstrip line dispersion results from the transmission mode conversion. At low frequencies, the signal is transmitted in the quasi-TEM mode and its

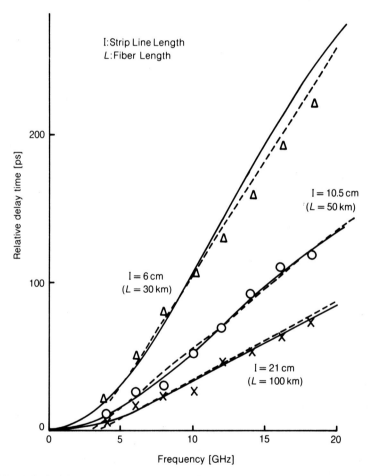

Fig. 3.35 Relative delay time characteristics of microstrip line equalizer. Solid lines show calculated values and broken lines show the required characteristics. △, ○ and × are measured values for three equalizers.
Source: Takachio and Iwashita (1988).

effective permittivity ε_{eff} is expressed as $(\varepsilon_r + 1)/2$, where ε_r is the relative permittivity of the substrate material. At high frequencies, the transmission mode is converted to the TE mode and the effective permittivity is ε_r. The delay time of dispersive media is expressed as:

$$\tau = \frac{l}{c}\left\{\sqrt{\varepsilon_{\text{eff}}} + f\frac{\text{d}}{\text{d}f}\sqrt{\varepsilon_{\text{reff}}}\right\} \tag{3.66}$$

where l is the microstrip line length and c is the velocity of light in free space.

The delay time is dependent upon frequency as the effective permittivity is dependent upon frequency. The delay time characteristics of microstrip lines are shown in Fig. 3.35. This equalizer can compensate dispersion from a 100 km single mode fiber with 21 cm length.

Chromatic dispersion compensation in heterodyne detection has already been shown under sinusoidal modulation (Takachio and Iwashita, 1988), CPFSK (Iwashita and Takachio, 1988) and PSK modulation transmission experiments (Takachio, Norimatsu and Iwashita, 1992).

The transmission experiments have been conducted by using a microstrip line delay equalizer. The experimental arrangement for delay equalization is shown in Fig. 3.36. The transmitter is a 1.55 μm three-electrode DFB laser diode. The fiber is a 202 km single mode fiber with 1.3 μm zero dispersion. The fiber chromatic dispersion is 16 ps/km/nm. The received IF signal is compensated by a 42 cm microstrip line delay equalizer and the frequency response of the microstrip line is compensated by an RC circuit. The error rate performance is shown in Fig. 3.37. Receiver sensitivity degradation due to transmission is completely compensated.

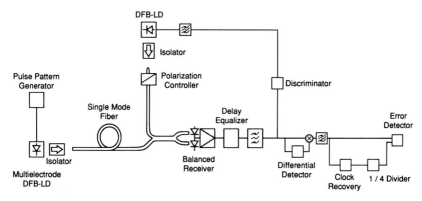

Fig. 3.36 Experimental arrangement for delay equalization at 8 Gbit/s 202 km single mode transmission.
Source: Iwashita and Takachio, 1990.

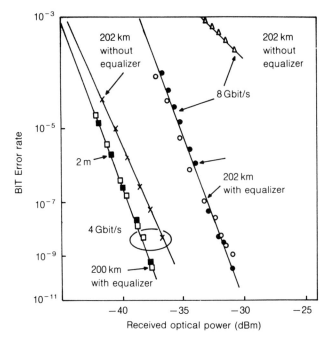

Fig. 3.37 Error rate performances for delay equalization with 4 and 8 Gbit/s CPFSK transmission elements.
Source: Iwashita and Takachio, 1990.

These compensation techniques will play an important role in multi-gigabit-per-second coherent transmission systems because they are easily applied to the heterodyne detection systems.

3.7 POLARIZATION COMPENSATION

Katsushi Iwashita

Coherent lightwave transmission systems require that the polarization of the received signal and local oscillator are matched. Signal polarization fluctuation, due to inherent birefringent properties of the fiber and those induced by thermal and mechanical stresses, presents an obstacle to coherent detection. The state of polarization in installed fibers varies slowly below 1 kHz but it varies endlessly (Imai and Matsumoto, 1988). Polarization-maintaining fibers as transmission media is one way to achieve a coherent transmission system as in Iwashita, *et al.* (1986). However, if we use a conventional single mode fiber, there are mainly three types of matching methods for this purpose as follows:

POLARIZATION COMPENSATION

1. *Polarization control*: controls the state of polarization of the received signal or local signal in a feedback loop as in Imai, Nosu and Yamaguchi (1985), Ulrich (1979), Matsumoto and Kano (1986) and Walker and Walker (1990).
2. *Polarization diversity*: the received signal is split into two orthogonal polarization components that are detected by two receivers; then the two electrical signals are combined as in Okoshi (1985), Imai (1991) and Shibutani and Yamazaki (1989).
3. *Polarization scrambling*: half of the signal power is transmitted in either of two orthogonal SOPs as in Cimini (1988).

3.7.1 Polarization control

Control methods of polarization are as follows:

1. Bulk half waveplate and quarter waveplate (Imai, Nosu and Yamaguchi, 1985).
2. Fiber type polarization controller (Ulrich, 1979, and Matsumoto and Kano, 1986).
3. Electro-optic effect (Walker and Walker, 1990).

The first method uses bulk type waveplates and rotates them to match the SOP. The second method uses birefringence induced by stresses. The third method utilizes electro-optic effects to vary SOP.

In the second method, fiber type polarization controllers utilize birefringence induced by stress. The stress is applied to the fiber by bending the fiber. Figure 3.38

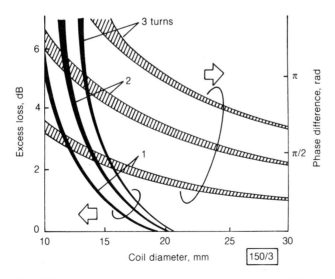

Fig. 3.38 Retardation insertion loss due to bending of fiber.
Source: Matsumoto and Kano, 1986.

shows the birefringence induced by bending the fiber as in Matsumoto and Kano (1986). Half waveplate, which has the birefringence characteristics of π rad phase shift, is achieved with 21 mm diameter and 3 turns. Quarter waveplate is achieved with 27 mm diameter and 2 turns.

A practical SOP controlling system must change the polarization endlessly to prevent momentary signal loss during the resetting of any constituent components. A schematic configuration of the endless rotatable fractional-wave device is shown in Fig. 3.39.

These polarization controlling methods are mainly based on mechanical change. Therefore, the response of the devices is important. To avoid problems, the following polarization diversity receiver is a promising method.

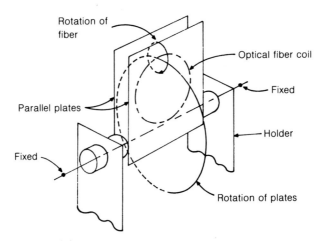

Fig. 3.39 Fiber polarization controller.
Source: Matsumoto and Kano, 1986.

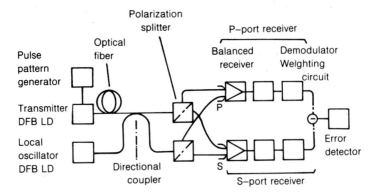

Fig. 3.40 Configuration of polarization diversity receiver.

3.7.2 Polarization diversity

The configuration of a polarization diversity receiver is shown in Fig. 3.40. The transmitted light and local oscillator light are split into two orthogonal components and detected by two receivers. The received signals are demodulated, and then they are combined according to the power level of each component. IF signal combination is possible but it is difficult to track endlessly the phase.

There are various methods to combine the two polarization state signals:

- maximum ratio combination;
- equal-gain combination;
- selection switch combination.

The receiver sensitivity degradation using these three methods for synchronous detection systems in shown in Fig. 3.41 according to Imai (1991). There is no sensitivity degradation in a maximum ratio combination. However, the case of a nonsynchronous detection system is different from a synchronous one. The sensitivity degradation in a polarization diversity receiver with nonsynchronous (square law detector) detection is shown in Fig 3.42 as in Imai (1991). In the case of the nonsynchronous detection, the sensitivity degradation from the ideal detection is about 0.3–0.4 dB.

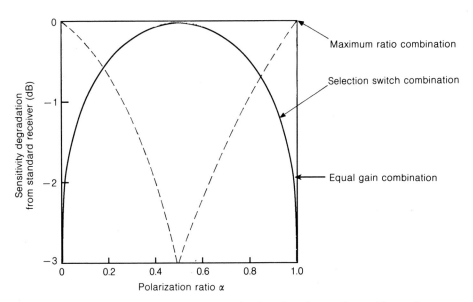

Fig. 3.41 Sensitivity degradation in a polarization diversity receiver with synchronous detection.
Source: Imai, 1991.

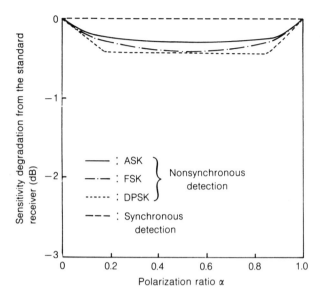

Fig. 3.42 Nonsynchronous detection.

3.8 HOMODYNE DETECTION

Katsushi Iwashita

Optical PSK homodyne detection offers the best sensitivity of any binary signalling technique and requires only the same electrical bandwidth as the bit rate. While it is suitable for multi-gigabit coherent transmission systems, it requires extremely narrow linewidth lasers. Laser linewidth requirements depend on the configuration of the phase-locked loop (PLL). A decision-driven PLL can tolerate a wider linewidth requirement, e.g. 6.2×10^{-4} times the bit rate as in Kazovsky (1985), than a balanced PLL which is 5.9×10^{-6} times the bit rate (Kazovsky, 1986). Therefore, the linewidth requirement can be tolerated by increasing the bit rate (Kahn, 1990). However, the linewidth requirement is also restricted by the PLL propagation delay time. Some papers have considered loop propagation delay theoretically as in Kazovsky (1985) and Norimatsu and Iwashita (1991). Here, we will calculate the requirement for spectral linewidth taking into consideration the PLL loop delay time.

3.8.1 PLL propagation time effect

The closed loop transfer function considering the loop delay time is expressed as follows:

$$H(s) = \frac{G(s\tau_2 + 1)e^{-s\tau}}{s^2 + G(s\tau_2 + 1)e^{-s\tau}} \tag{3.67}$$

HOMODYNE DETECTION

By substituting equation 3.67 with 2.55 and 2.56, the phase noise and shot noise variance can be obtained. The allowable beat linewidth considering the loop delay time was obtained approximately as follows by Norimatsu and Iwashita (1991).

$$\sigma_{PN2}^2 = \frac{36 + 36y - 30y^2 + 6y^3 - y^4}{36 - 72y + 6y^2 - y^4} \sigma_{PN0}^2 \qquad (3.68)$$

$$\sigma_{SN2}^2 = \frac{36 + 12y - 18y^2 + 6y^3 - y^4}{36 - 72y + 6y^2 - y^4} \sigma_{SN0}^2 \qquad (3.69)$$

where

$$\sigma_{PN0}^2 = \frac{\pi \Delta \nu}{G \tau_2}, \quad \sigma_{SN0}^2 = \frac{e}{4Rk_s P_s} \frac{x+1}{2\tau_2}, \quad x = G\tau_2^2, \quad y = \frac{\tau}{\tau_2}$$

and k_s is the received power splitting ratio to the Q-arm.

The case when the damping factor ζ is $1/\sqrt{2}$ is considered. Therefore, the expression of the phase-error variance is a sum of σ_{PN2}^2 and σ_{SN2}^2. The standard deviation of phase error σ with respect to the loop natural frequency ω_n for several loop delay times τ is shown in Fig. 3.43. The minimized standard deviation of phase error increases gradually as the loop delay time increases.

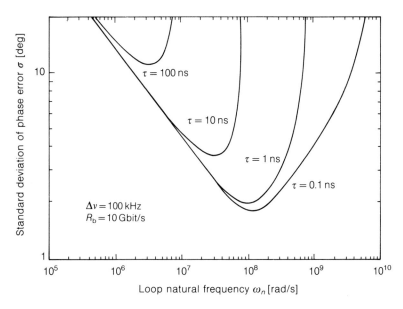

Fig. 3.43 Standard deviation of phase error σ versus loop natural frequency ω_n for several loop delay times.
Source: Norimatsu and Iwashita, 1991.

The normalized permitted maximum beat linewidth with respect to the normalized loop delay time is plotted in Fig. 3.44. When $\tau = 0$, the required beat linewidth corresponds to the results given in Kazovsky (1985). The propagation time, greater than $1/R_b$, is not negligible. When the loop delay time τ is not negligible, the required beat linewidth Δv is not proportional to the bit rate but limited by the condition of:

$$\Delta v = 2.04 \times 10^{-3}/\tau \tag{3.70}$$

This condition gives a 1 dB power penalty.

The power penalty under the nonnegligible loop delay time τ is investigated. The loop delay time τ with fixed linewidth Δv and the beat linewidth Δv with fixed loop delay time τ are changed in the vicinity of $\Delta v/R_b = 1.02 \times 10^{-6}$ or $\tau R_b = 2 \times 10^3$ in Fig. 3.44. This result is shown in Fig. 3.45.

A homodyne detection transmission experiment using laser diodes was successfully conducted. The experimental setup is shown in Fig. 3.46 according

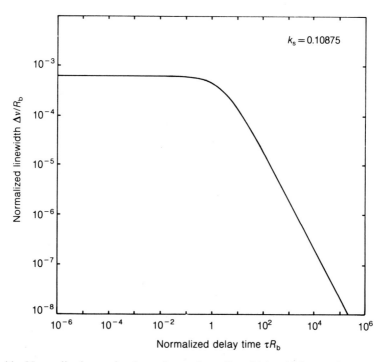

Fig. 3.44 Normalized permitted maximum beat linewidth which permits 1 dB power penalty at a bit error rate of 10^{-10} with respect to normalized loop delay time. Source: Norimatsu and Iwashita, 1991.

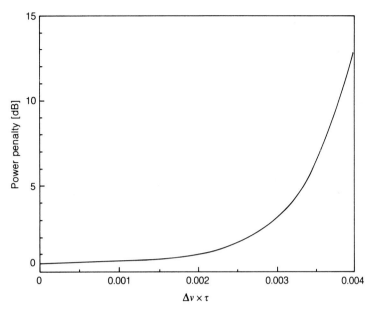

Fig. 3.45 Power penalty from ideal detection due to $\Delta v \tau$ under significant influence of loop delay time.
Source: Norimatsu and Iwashita, 1991.

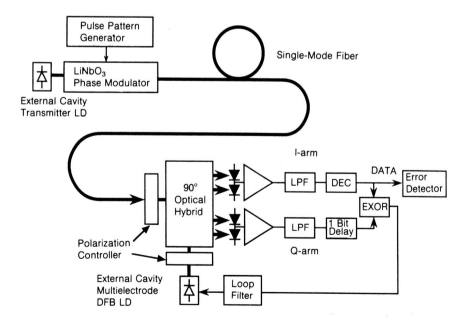

Fig. 3.46 Experimental setup of PSK homodyne detection experiment.

to Norimatsu and Iwashita (1991b). The transmitter and the local oscillator are 1.55 μm wavelength DFB laser diodes with external cavities. Fiber-pigtailed travelling-wave electrode Ti:LiNbO$_3$ optical phase modulators are used to produce 10 Gbit/s PSK signals. The two transmitter signals are combined after adjusting their polarization states. They are amplified with a backward pumped Er-doped fiber amplifier. The amplified signals are transmitted through a dispersion-shifted fiber with 1.554 μm zero dispersion and 0.21 dB/km loss.

The optical 90° hybrid is composed of two polarization controllers, a 1:1 half mirror and two polarization beam splitters. The four outputs of the 90° hybrid are detected by two dual-pin detectors. The optical phase is recovered by a decision-driven phase-locked loop. The control signal is passed through the loop filter and applied to a laser diode current. To enhance the bandwidth of the phase-locked loop, a multi-electrode DFB laser diode with external cavity is used as an optical source for the local oscillator. The beat linewidth is 150 kHz and the loop delay time is 13 ns.

The error rate performance of 10 Gbit/s PSK homodyne detection with and without transmission using an Er-doped optical post amplifier is shown in Fig. 3.47. The fiber input power is +12.5 dBm. The receiver sensitivity at the error rate of 10^{-9} is -38.1 dBm (121 photons/bit). There is no degradation even with a 231 km transmission span.

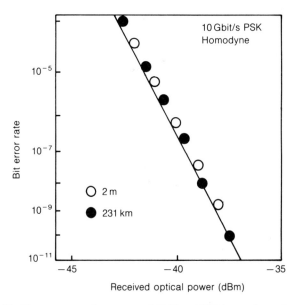

Fig. 3.47 Error rate performance of 230 km PSK homodyne detection. Source: Iwashita and Norimatsu, 1991.

REFERENCES

Agrawal, G. P. (1989) *Nonlinear Fiber Optics*, Academic Press, London.

Ainsle, B. J., Beales, K. J., Cooper, D. M., Day, C. R. and Rush, J. D. (1982) Monomode fiber with ultra-low loss and minimum dispersion at 1.55 m, *Electronics Letters*, **18**, pp. 842–844.

Anderkson, P. A. (1991) Picosecond optical sampling using four-wave mixing in fiber, *Electronics Letters*, **27**, pp. 1440–1441.

Anderkson, P. A., Olsson, N. A., Simpson, J. R., Tanbun-Ek, T., Rogan, R. A. and Haner, M. (1991) 16 Gbit/s all optical demultiplexing using four-wave mixing, *Electronics Letters*, **27**, pp. 922–924.

Aoki, Y., Tajima, K. and Mito, I. (1988) Input power limits of single-mode optical fibers due to stimulated Brillouin scattering in optical communication systems, *Journal Lightwave Technology*, **6**, pp. 710–719.

Azuma, Y., Shibata, N., Horiguchi, T. and Tateda, M. (1988) Wavelength dependence of Brillouin-gain spectra for single-mode optical fibers, *Electronics Letters*, **24**, pp. 251–252.

Bergano, N., Poole, C. D. and Wagner, R. E. (1987) Investigation of polarization dispersion in long lengths of single-mode fiber using multilongitudinal mode lasers, *Journal Lightwave Technology*, **LT-5**, pp. 1618–1622.

Bhagavatula, V. A., Spotz, M. S. and Love, W. F. (1984) Dispersion-shifted segmented-core single-mode fibers, *Optics Letters*, **9**, pp. 186–188.

Bolle, A., Grosso, G. and Daino, B. (1989) Brillouin gain curve dependence on frequency spectrum of PSK-modulated signals, *Electronics Letters*, **25**, pp. 2–3.

Chraplyvy, A. R. (1984) Optical power limits in multichannel wavelength-division-multiplexed systems due to stimulated Raman scattering, *Electronics Letters*, **20**, pp. 58–59.

Chraplyvy, A. R. (1990) Limitations on lightwave communications imposed by optical-fiber nonlinearities, *Journal Lightwave Technology*, **8**, pp. 1548–1557.

Chraplyvy, A. R. and Henry, P. S. (1983) Performance degradation due to stimulated Raman scattering in wavelength-division-multiplexed optical fibre systems, *Electronics Letters*, **19**, pp. 641–642.

Chraplyvy, A. R. and Stone, J. (1984) Measurement of crossphase modulation in coherent wavelength-division multiplexing using injection lasers, *Electronics Letters*, **20**, pp. 996–997.

Cimini, L. J. (1988) Polarization-insensitive coherent lightwave system using wide-deviation FSK and data-induced polarization switching, *Electronics Letters*, **24/6**, pp. 358–360.

Cohen, L. G. (1985) Comparison of single-mode fiber dispersion measurement techniques, *Journal Lightwave Technology*, **LT-3**, pp. 958–966.

Cohen, L. G. and Lin, C. (1978) Universal fiber-optic (UFO) measurement system based on a near-IR fiber Raman laser, *IEEE Journal Quantum Electronics*, **QE-14**, pp. 855–859.

Cotter, D. (1983) Stimulated Brillouin scattering in monomode optical fiber, *Journal Optical Communications*, **4**, pp. 10–19.

Daikoku, K. and Sugimura, A. (1978) Direct measurement of wavelength dispersion in optical fibers – difference method, *Electronics Letters*, **14**, pp. 149–151.

Davis, A. W., Pettitt, M. J., King, J. P. and Wright, S. (1987) Phase diversity techniques for coherent optical receivers, *Journal Lightwave Technology*, **LT-5/4**, pp. 561–572.

Elrafaie, A. F., Wagner, R. E., Atlas, D. A. and Daut, D. G. (1988) Chromatic dispersion limitations in coherent lightwave transmission systems, *Journal Lightwave Technology*, **6/5**, pp. 704–709.

Glance, B. S. (1986) An optical heterodyne mixer providing image-frequency rejection, *IEEE Journal of Lightwave Technology*, **LT–4/11**, pp. 1722–1725.

Gloge, D. (1971) Dispersion in weakly guiding fibers, *Applied Optical*, **10**, pp. 2442–2445.

Heiman, D., Hamilton, D. S. and Hellwarth, R. W. (1979) Brillouin scattering measurements on optical glasses, *Physical Review*, **B19**, pp. 6583–6592.

Henry, C. H. (1982) Theory of the linewidth of semiconductor lasers, *IEEE Journal Quantum Electronics*, **QE-18/2**, pp. 259–264.

Hill, K. O., Johnson, D. C., Kawasaki, B. S. and MacDonald, R. I. (1978) CW three-wave mixing in single-mode opticals fibers, *Journal Applied Physics*, **49**, pp. 5098–5106.

Hosaka, T., Okamoto, K., Miya, T., Sasaki, Y. and Edahiro, T. (1981) Low-loss single-polarisation fibers with asymmetrical strain birefringence, *Electronics Letters*, **17**, pp. 530–531.

Hussey, C. D. and Pask, C. (1982) Theory of the profile-moments description of single-mode fibers, *IEE Proceedings*, **129**, PT-H, pp. 123–134.

Imai, T. (1991) Sensitivity degradation in polarization diversity receivers for lightwave systems, *Journal Lightwave Technology*, **9**, pp. 650–658.

Imai, T. and Matsumoto, T. (1988) Polarization fluctuation in a single-mode optical fiber, *Journal Lightwave Technology*, **LT–6/9**, pp. 1366–1375.

Imai, T., Nosu, K. and Yamaguchi, E. (1985) Optical polarization control utilizing an optical heterodyne detection scheme, *Electronics Letters*, **21/2**, pp. 52–53.

Inoue, K. and Toba, H. (1992) Wavelength conversion experiment using fiber four-wave mixing, *IEEE Photonics Technology Letters*, **4**, pp. 69–72.

Ippen, E. P. and Stolen, R. H. (1972) Stimulated Brillouin scattering in optical fibers, *Applied Physics Letters*, **21**, pp. 539–541.

Ishida, O., Toba, H. and Tohmori, Y. (1989) Pure frequency modulation of a multielectrode distributed-Bragg-reflector (DBR) laser diode, *IEEE Photonics Technology Letters*, **1/7**, pp. 156–158.

Iwashita, K. and Matsumoto, T. (1987) Modulation and detection characteristics of optical continuous phase FSK transmission systems, *Journal Lightwave Technology*, **LT–5/4**, pp. 452–460.

Iwashita, K. and Norimatsu, S. (1991) Cross-phase modulation influence on a two channel optical PSK homodyne transmission system. *ECOC'91*, pp. 661–664.

Iwashita, K. and Takachio, N. (1988) Compensation of 202 km single-mode fibre chromatic dispersion in 4 Gb/s optical CPFSK transmission experiment, *Electronics Letters*, **24**, pp. 759–760.

Iwashita, K. and Takachio, N. (1989) Experimental evaluation of coherent dispersion distortion in optical CPFSK transmission systems, *Journal Lightwave Technology*, **7/10**, pp. 1484–1487.

Iwashita, K. and Takachio, N. (1990) Chromatic dispersion compensation in coherent optical communications. *Journal Lightwave Technology*, **8**(3), pp. 367–375.

Iwashita, K., Kano, H., Matsumoto, T, and Sasaki, Y. (1986) FSK transmission experiment using 10.5 km polarization-maintaining fibre, *Electronics Letters*, **22/4**, pp. 214–215.

Jen, C. K., Oliveira, J. E. B., Goto, N. and Abe, K. (1988) Role of guided acoustic wave properties in single-mode optical fibre design, *Electronics Letters*, **24**, pp. 1419–1420.

Jen, C. K., Safaai-Jazi, A. and Farnell, G. W. (1986) Analysis of weakly guiding fiber acoustic waveguides, *IEEE Transactions Ultrasonics, Ferroelectric Frequency Contr.*, **UFFC–33**, pp. 634–643.

Kahn, J. M. (1990) BPSK homodyne detection experiment using balanced optical phase-locked loop with quantized feedback, *IEEE Photonics Technology Letters*, **2/11**, pp. 840–843.

Kaminow, I. P. (1981) Polarization in optical fibers, *IEEE Journal Quantum Electronics*, **QE-17**, pp. 15–22.

Katsuyama, T., Matsumura, M. and Suganuma, T. (1981) Low-loss single-polarization fibres, *Electronics Letters*, **17**, pp. 473–474.

Kawakami, S. and Ikeda, M. (1978) Transmission characteristics of a two mode optical waveguide, *IEEE Journal Quantum Electronics*, **QE-14**, pp. 608–614.

Kawano, K., Kitoh, T., Mitomi, O., Nozawa, T. and Jumonji, H. (1989) A wide-band and low-driving-power phase modulator employing a Ti:LiNbO$_3$ optical waveguide at 1.5 mm wavelength, *IEEE Photonics Technology Letters*, **1/2**, pp. 33–34.

Kazovsky, L. G. (1985) Decision-driven phase-locked loop for optical homodyne receivers: performance analysis and laser linewidth requirements, *Journal Lightwave Technology*, **LT-3/6**, pp. 1238–1247.

Kazovsky, L. G. (1986) Balanced phase-locked loops for homodyne receivers: performance analysis, design considerations, and laser linewidth requirements, *Journal Lightwave Technology*, **LT-4/2**, pp. 182–195, Feb.

Kikuchi, K. (1989) Effect of 1/f-type FM noise on semiconductor laser linewidth residual in high-power limit, *IEEE Journal Quantum Electronics*, **QE-25/4**, pp. 684–688.

Kimura, T. and Sugimura, A. (1987) Linewidth reduction by coupled phase-shift distributed-feedback lasers, *Electronics Letters*, **23**, pp. 1014–1015.

Kobayashi, K. and Mito, I. (1988) Single frequency and tunable laser diodes, *Journal Lightwave Technology*, **6/11**, pp. 1623–1633.

Kobayashi, S., Shibata, S., Shibata, N. and Izawa, T. (1981) Wavelength dispersion characteristics of single-mode fibers in low-loss region, *IEEE Journal Quantum Electronics*, **QE-16**, pp. 215–225.

Lichtman, E., Waarts, R. G. and Friesem, A. A. (1989) Stimulated Brillouin scattering excited by a modulated pump wave in single-mode fibers, *Journal Lightwave Technology*, **7**, pp. 171–174.

Linke, R. A. and Gnauck, A. H. (1988) High-capacity coherent lightwave systems, *Journal Lightwave Technology*, **6**, pp. 1750–1769.

Maritoson, I. H. (1965) Interspecimen comparison of the refractive index of fused silica, *Journal Optical Society America*, **55**, pp. 1205–1209.

Matsui, Y., Kunii, T., Horikawa, H. and Kamijoh, T. (1991) Narrow-linewidth (< 200 kHz) operation of 1.5 µm Butt-jointed multiple-quantum well-distributed Bragg reflector laser, *IEEE Photonics Technology Letters*, **3/5**, pp. 424–426.

Matsumoto, T. and Kano, H. (1986) Endlessly rotatable fractional-wave devices for single-mode-fibre optics, *Electronics Letters*, **22/2**, pp. 78–79.

Monerie, M., Lamonir, P. and Veunhomme, L. (1980) Polarisation mode dispersion in long single-mode fibres, *Electronics Letters*, **16**, pp. 907–908.

Murata, S., Mito, I. and Kobayashi, K. (1987) Spectral characteristics for a 1.5 µm DBR laser with frequency-tuning region, *IEEE Journal Quantum Electronics*, **QE-23/6**, pp. 835–838.

Norimatsu, S. and Iwashita, K. (1991a) Cross-phase modulation influence on a two-channel optical PSK homodyne transmission system, *IEEE Photonics Technology Letters*, **3**, pp. 1142–1144.

Norimatsu, S. and Iwashita, K. (1991b) PLL propagation delay-time influence on linewidth requirements of optical PSK homodyne detection, *Journal Lightwave Technology*, **9/10**, pp. 1367–1375.

Norimatsu, S., Iwashita, K. and Noguchi, K. (1990) 10 Gbit/s optical PSK homodyne transmission experiments using external cavity DFB LDs, *Electronics Letters*, **26/10**. pp. 648–649.

Nosu, K. and Iwashita, K. (1988) A consideration on factors affecting future coherent lightwave communications, *Journal Lightwave Technology*, **LT–6/5**, pp. 686–694.

Ohashi, M., Kuwaki, N., Tanaka, C., Uesugi, N. and Negishi, Y. (1986) Bend-optimized dispersion-shifted step-shaped-index (SS) fibres, *Electronics Letters*, **21**, pp. 1285–1286.

Ohashi, M., Kuwaki, N. and Uesugi, N. (1987) Characteristics of dispersion-shifted fibers, *Review ECL*, **35**, pp. 535–539.

Ohkawa, N. (1988) Fiber-optic multigigabit GaAs MIC front-end circuit with inductor peaking, *Journal Lightwave Technology*, **6/11**, pp. 1665–1671.

Okai, M., Tsuchiya, T., Uomi, K., Chinone, N. and Harada, T. (1990) Corrugation-pitch-modulated MQW-DFB laser with narrow spectral linewidth (170 kHz), *IEEE Photonics Technology Letters*, **2/8**, pp. 529–530.

Okamoto, K., Edahiro, T. and Shibata, N. (1982) Polarization properties of single-polarization fibers, *Optics Letters*, **7**, pp. 569–571.

Okamoto, K., Hosaka, T. and Sasaki, Y. (1982) Linear single polarization fibers with zero polarization mode dispersion, *IEEE Journal Quantum Electronics*, **QE–18**, pp. 496–503.

Okoshi, T. (1985) Polarization-state control schemes for heterodyne or homodyne optical fiber communications, *Journal Lightwave Technology*, **LT–3/6**, pp. 1232–1237.

Personick, S. D. (1971) Time dispersion in dielectric waveguides, *Bell Systems Technical Journal*, **50**, pp. 843–858.

Personick, S. D. (1973) Receiver design for digital optical communication systems, Part I and II, *Bell Systems Technical Journal*, **52**, pp. 843–886.

Petermann, K. (1983) Constrains for fundamental mode spot size for broad-band dispersion-compensated single-mode fibres, *Electronics Letters*, **19**, pp. 712–714.

Poole, C. D. (1989) Measurement of polarization-mode dispersion in single-mode fibers with random mode coupling, *Optics Letters*, **10**, pp. 523–525.

Priest, R. G. and Giallorenzi, T. G. (1987) Dispersion compensation in coherent fiber-optic communications, *Optics Letters*, **1/8**, pp. 622–624.

Ramaswamy, V., Kaminow, I. P. and Kaiser, P. (1978) Single polarization optical fibers: exposed cladding technique, *Applied Physics Letters*, **33**, pp. 814–816.

Rashleigh, S. and Ulrich, R. (1978) Polarization mode dispersion in single-mode fibers, *Optics Letters*, **3**, pp. 60–62.

Saito, S., Nilsson, O. and Yamamoto, Y. (1985) Frequency modulation noise and linewidth reduction in a semiconductor laser by means of negative frequency feedback semiconductor laser, *Applied Physics Letters*, **46/1**, pp. 3–5, Jan 1.

Shen, Y. R. (1984) *The Principles of Nonlinear Optics*, John Wiley, New York.

Shibata, N., Azuma, Y., Tateda, M. and Nakano, Y. (1988) Experimental verification of efficiency of wave generation through four-wave mixing in low-loss dispersion-shifted single-mode optical fibre, *Electronics Letters*, **24**, pp. 1528–1529.

Shibata, N., Azuma, Y., Horiguchi, T. and Tateda, M. (1988) Identification of longitudinal acoustic modes guided in the core region of a single-mode optical fiber by Brillouin gain spectra measurements, *Optics Letters*, **13**, pp. 595–597.

Shibata, N., Braun, R. P. and Waarts, R. G. (1986) Crosstalk due to three-wave mixing process in a coherent single-mode transmission line, *Electronics Letters*, **22**, pp. 675–677.

Shibata, N., Braun, R. P. and Waarts, R. G. (1987) Phase-mismatch dependence of efficiency of wave generation through four-wave mixing in a single-mode optical fiber, *IEEE Journal Quantum Electronics*, **QE–23**, pp. 1205–1210.

Shibata, N., Nosu, K., Iwashita, K. and Azuma, Y. (1990) Transmission limitations due to fiber nonlinearities in optical FDM systems, *IEEE Journal Select. Areas Communications*, **8**, pp. 1068–1077.

Shibata, N., Okamoto, K. and Azuma, Y. (1989) Longitudinal acoustic modes and Brillouin gain spectra for GeO_2-doped doped-core single-mode fibers, *Journal Optical Society America*, **B/6**, pp. 1167–1174.

Shibata, N., Okamoto, K., Suzuki, K. and Ishida, Y. (1983) Polarization-mode properties of elliptical-core fibers and stress-induced birefringent fibers, *Journal Optical Society America*, **73**, pp. 1792–1798.

Shibata, N., Tateda, M. and Seika, S. (1982) Polarization mode dispersion measurement in elliptical core single-mode fibers by a spatial technique, *IEEE Journal Quantum Electronics*, **QE–18**, pp. 53–58.

Shibata, N., Tsubokawa, M. and Seikai, S. (1984) Measurements of polarization mode dispersion by optical heterodyne detection, *Electronics Letters*, **20**, pp. 1055–1057.

Shibata, N., Uchida, N., Tateda, M. and Seikai, S. (1982) Normalised frequency dependence of polarisation mode dispersion due to thermal-stress-induced birefringence in an elliptical core single-mode fibre, *Electronics Letters*, **18**, pp. 563–564.

Shibata, N., Waarts, R. G. and Braun, R. P. (1987) Brillouin gain spectra for single-mode fibers having pure-silica, GeO_2-doped and P_2O_5-doped cores, *Optics Letters*, **12**, pp. 269–271.

Shibutani, M. and Yamazaki, S. (1989) A study on an active square-law combining method for a polarization-diversity coherent optical receiver, *IEEE Photonics Technology Letters*, **1/7**, pp. 182–183.

Sinha, N. K. (1978) Normalized dispersion of birefringence of quartz and stress-optical coefficient of fused silica and plate glass, *Physics and Chemistry of Glasses*, **19**, pp. 67–77.

Smith, R. G. (1972) Optical power handling capacity of low loss optical fibers as determined by stimulated Raman and Brillouin scattering, *Applied Optics*, **11**, pp. 2489–2494.

Stolen, R. H. (1979) Nonlinear properties of optical fibers, in *Optical Fiber Telecommunications*, Miller, S. E. and Chynoweth, A. G., Eds., Academic Press, New York.

Sugie, T. (1991a) Suppression of SBS by discontinuous Brillouin frequency shifted fibre in CPFSK coherent lightwave system with booster amplifier, *Electronics Letters*, **27**, pp. 1231–1232.

Sugie, T. (1991b) Transmission limitations of CPFSK coherent lightwave systems due to stimulated Brillouin scattering in optical fiber, *Journal Lightwave Technology*, **9**, pp. 1145–1155.

Sunde, E. D. (1961) Pulse transmission by AM, FM, and PM in the presence of phase distortion, *Bell System Technical Journal*, **40**, p. 353.

Takachio, N. and Iwashita, K. (1988) Compensation of fiber chromatic dispersion in optical heterodyne detection, *Electronics Letters*, **24/2**, pp. 108–109.

Takachio, N., Iwashita, K., Hata, S., Onadera, K., Katsura, K. and Kikuchi, H. (1990) A 10 Gb/s optical heterodyne detection experiment using a 23 GHz bandwidth balanced receiver, *IEEE Transactions Microwave Theory Technology*, **38**, pp. 1900–1905.

Takachio, N., Norimatsu, S. and Iwashita, K. (1992) Optical PSK synchronous heterodyne detection transmission experiment using fiber chromatic dispersion equalisation, *IEEE Photon Technology Letters*, **4/3**, pp. 278–280.

Tateda, M., Shibata, N. and Seikai, S. (1981) Interferometric method for chromatic dispersion in a single-mode optical fiber, *IEEE Journal Quantum Electronics*, **QE-17**, pp. 404–407.

Thomas, P. J., Rowell, N. L., van Driel, H. M. and Stegeman, G. I. (1979) Normal acoustic modes and Brillouin scattering in single-mode optical fiber, *Physical Review*, **B19**, pp. 4986–4998.

Tkach, R. W., Chraplyvy, A. R. and Derosier, R. M. (1986) Spontaneous Brillouin scattering for single-mode optical fibre characterisation, *Electronics Letters*, **22**, pp. 1011–1013.

Tsubokawa, M. and Sasaki, Y. (1988) Limitation of transmission distance and capacity due to polarisation dispersion in a lightwave system, *Electronics Letters*, **24**, pp. 350–352.

Uesugi, N., Ikeda, M. and Sasaki, Y. (1981) Maximum single-frequency input power in a long optical fiber determined by stimulated Brillouin scattering, *Electronics Letters*, **17**, pp. 379–380.

Ulrich, R. (1979) Polarization stabilization on single-mode fiber, *Applied Physics Letters*, **35/11**, pp. 840–842.

Vobian, J. (1990) Chromatic and polarization dispersion measurements of single-mode fibers with a Mach-Zehnder interferometer between 1200 and 1700 nm, *Journal Optical Communications*, **11**, pp. 29–36.

Walker, N. G. and Walker, G. R. (1990) Polarization control for coherent communications, *Journal Lightwave Technology*, **8**, pp. 438–458.

Winter, J. H. (1989) Equalization in coherent lightwave systems using microwave waveguides, *Journal Lightwave Technology*, **7**, pp. 813–815.

Wyatt, R. and Devlin, W. J. (1983) 10 kHz 1.5 µm InGaAsP external cavity laser with 55 nm tuning range, *Electronics Letters*, **19**, pp. 110–112.

Yamazaki, S., Kimura, K., Shikada, M., Yamaguchi, M. and Mito, I. (1985) Realization of flat FM response by directly modulating a phase tunable DFB laser diode, *Electronics Letters*, **21/7**, pp. 283–285, Mar. 28.

Yasaka, H., Fukuda, M. and Ikegami, T. (1988) Current tailoring for lowering linewidth floor, *Electronics Letters*, **24/12**, pp. 760–762.

Yoshikuni, Y. and Motosugi, G. (1986) Independent modulation in amplitude and frequency regimes by a multielectrode distributed-feed-back laser, *9th Conference Optical Fiber Communications*, Atlanta, GA, Feb. 24–26, paper TuF1.

4
Optical filters and couplers

4.1 OPTICAL FILTERS

Hiromu Toba

Optical filters or optical multiplexers/demultiplexers are indispensable for the construction of WDM or optical FDM systems as in Ishio, Minowa and Nosu (1984) and Nosu, Toba and Iwashita (1987). The advent of coherent light sources has enabled us to densely pack optical carriers in the optical frequency region as is possible in radio communication systems. In order to multi/demultiplex narrowly spaced optical signals, we need filters with high resolution. Some filter configurations developed for microwave and millimeter-wave systems can be applied to optical FDM systems as in Ohtomo, Shimada and Suzuki (1971).

This section describes the configuration of optical filters and design of waveguide type optical filters. Tunable optical filters are characterized in 4.1.3.

4.1.1 Configuration of optical filters

Figure 4.1 classifies optical filters based on their filtering principle. There are three type of filters:

- two-beam interference;
- resonance;
- multiple beam interference.

Directional coupler type filters (Takato, *et al.*, 1988) and Mach-Zehnder interferometer filters (MZ filters) (Inoue, *et al.*, 1988) are based on two-beam interference. The transmittance of these filters shows a sinusoidal dependency on the optical frequency or wavelength. Channel spacing of a directional coupler type filter can be as large as 100 nm, while that of the MZ filter ranges from 0.01 nm to 100 nm according to Takato, *et al.* (1990a). These devices require precise control of the waveguide path length, that is the coupling region length for directional coupler filters and arm length difference for the MZ filters. They are normally fabricated as fiber type or planar waveguide type devices. The basic configuration of these

Principle	Filter	Configuration
two-beam interference	directional coupler	
	MZ filter	
resonance	interference thin film filter	
	FP filter	
	ring resonator	
	Bragg grating filter	
multi-beam interference	difrection grating	λ_1, λ_2 λ_1 λ_2
	arrayed-waveguide grating	
	transversal filter	

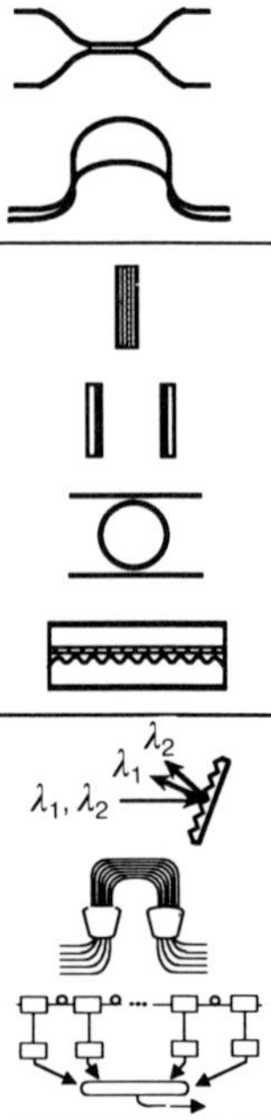

Fig. 4.1 Optical filters for OFDM/WDM systems.

filters is 2 by 2, so a serial or cascaded connection is needed for multi/demultiplexing more than three wavelengths.

Interference thin-film filters, Fabry-Perot interferometer filters (FP filters), ring resonators, and a Bragg reflection grating are based on resonance. Interference thin-film filters are fabricated by the serial deposition of two dielectric materials as in Minowa and Fujii (1983). A multi/demultiplexer utilizing this filter is fabricated as a bulk type device, and employs collimating and focusing lenses and an optical beam passing through the filter and/or is reflected by the filter (Ishio, Minowa and Nosu, 1984). This type of multi/demultiplexer is available for a small number of channels. Commercialized WDM systems that have two channels utilize this type of multi/demultiplexer according to Kanada, et al. (1983).

FP filters and ring resonators have a Lorentzian transmittance spectrum. Wavelength selectivity is determined by the Q-value of the resonator. FP filters are either bulk as in Frenkel and Lin (1989) or fiber type devices as in Stone and Stulz (1987), while ring resonators are fabricated as planar waveguides as in Ohmori, et al. (1990). The Bragg reflection grating filter is a waveguide with a surface corrugation grating and it reflects the wavelength, which satisfies the Bragg condition of the grating (Flanders, et al., 1974). The wavelength pass bandwidth ranges from sub-nanometers to several tens of nanometers according to Henry, et al. (1989).

The diffraction grating type filter is based on multiple interference; the diffracted angle changes with the signal wavelength. To date, a 1 nm-spaced, 32-channel optical multi/demultiplexer has been fabricated as a bulk type device (Wisely, 1991).

Arrayed waveguide grating filters and transversal filters are fabricated as planar waveguides. Arrayed waveguide filters consist of input/output waveguides, two focusing fan-shaped slab waveguide regions and a phase-array region of multiple channel waveguides with an optical path length difference ΔL between any two adjacent waveguides as in Takahashi, et al. (1990) and Dragone, Edwards and Kistler (1991). The light beam from the input waveguides radiates in the fan-shaped slab waveguide and then couples into the arrayed channel waveguides. After traveling through the arrayed waveguides, the light converges on the focal plane, where the ends of the output waveguides are located. The relation between the focal position x and frequency f is given by:

$$\frac{dx}{df} = \frac{Fm\lambda^2}{n_s dc} \quad (4.1)$$

where

$$m = \frac{n_c \Delta L}{\lambda}$$

where F is the focal length of the fan-shaped slab waveguide, n_c the effective refractive index of the channel waveguide, n_s the effective refractive index of the slab waveguide, d the pitch of the arrayed waveguide end and m the refraction order as in Takahashi, Suzuki and Nishi (1991). Channel frequency spacing of the filter is determined by the equation. A 10 GHz-spaced 11-channel multi/demultiplexer was fabricated for an SiO_2–GeO_2/SiO_2 waveguide at 1.55 μm with $F = 5762$ mm, $d = 20$ mm and $\Delta L = 1656$ mm. It corresponds to the linear dispersion $dx/df = 25$ μm/10 GHz, and $m = 1550$ as in Takahashi, Nisi and Hibino (1992). The number of arrayed waveguides was 41, which is sufficient to receive all the power radiating from the input power. This type of multi/demultiplexer has the advantage that all channels are simultaneously multi/demultiplexed and offers wavelength-selective $N \times N$ correction.

An optical transversal filter consists of variable power dividers, delay lines (taps) and phase shifters (Sasayama, Okuno and Habara, 1989). The arbitrary frequency characteristics of:

$$H(\omega) = \sum_{0}^{N-1} a_k \exp(-jk\omega\tau) \quad (4.2)$$

can be produced by tuning tap coefficients a_k, where N is the number of taps and τ the unit delay time caused by the optical delay line. Arbitrary complex tap coefficient a_k can be expressed by controlling the electric field and phase of optical signals. A 16-tap transversal filter with unit delay length of 10 mm ($\tau = 50$ ps) has been fabricated using a silica waveguide. This filter can select arbitrary channels out of 2.5 GHz equally spaced eight channels according to Sasayama, Okuno and Habara (1992).

4.1.2 Design of waveguide-type optical filter

Some waveguide filters employ the frequency dependence of the phase change by utilizing path length differences. They are similar to the multi/demultiplexers, investigated in the field of microwave and millimeter-wave communication technologies. Configurations of typical waveguide-type filters and their transmittance are shown in Fig. 4.2.

Figure 4.2(a) shows a Mach-Zehnder interferometer filter (MZ filter), which is a type of "periodic filter" as defined in the microwave and millimeter-wave communication technology fields as in Inoue, et al. (1988). The lengths of the two waveguide "arms" differ slightly from each other and an optical phase difference occurs at the end of the arms. Transmittance from port 1 to 3, T_{1-3}, and port 1 to 4, T_{1-4}, is described by:

$$T_{1-3} = \cos^2\left(\frac{\phi}{2}\right) \tag{4.3}$$

$$T_{1-4} = \sin^2\left(\frac{\phi}{2}\right) \tag{4.4}$$

where:

$$\phi = \frac{2\pi n \Delta L f}{c} \tag{4.5}$$

n is the effective refractive index of the waveguide, ΔL is the waveguide length difference between the two arms, f is the optical frequency, and c is the light velocity in a vacuum. The frequency spacing Δf of this filter is:

$$\Delta f = \frac{c}{2n\Delta L} \tag{4.6}$$

For example, a waveguide length difference $\Delta L = 10$ mm corresponds to the frequency spacing $\Delta f = 10$ GHz for silica waveguides, where $n = 1.45$, and the wavelength is 1.55 μm ($f = 193$ THz).

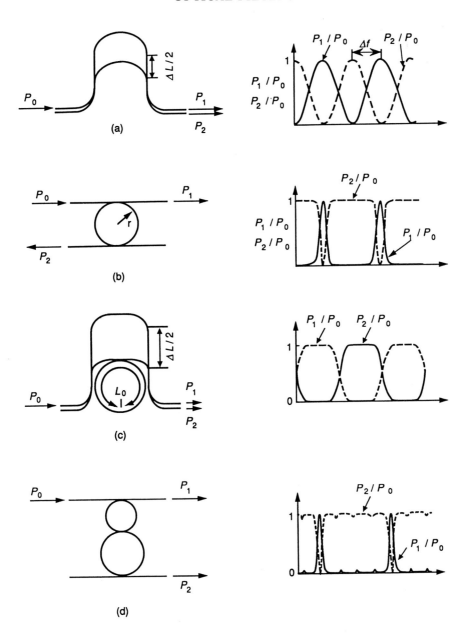

Fig. 4.2 Configurations of waveguide-type optical filters: (a) MZ filter (periodic filter); (b) ring resonator; (c) MZ filter with ring resonator; (d) double cavity ring resonator.

Figure 4.2(b) shows a ring resonator according to Ohmori, et al. (1990). Transmittance of the ring resonator is written as:

$$T_{1-2} = \frac{T_r^2}{(1-R_r)^2 + 4R_r \sin^2(\beta\pi r)} \qquad (4.7)$$

where:

$$T_r = k\exp(-\alpha\pi r) \qquad \text{(equivalent transmittance)} \qquad (4.8)$$

$$R_r = (1-k^2)\exp(-2\alpha\pi r) \qquad \text{(equivalent reflectance)} \qquad (4.9)$$

r is the ring radius, k is the amplitude coupling efficiency of the ring resonator, β is the propagation constant, and α is the attenuation constant. The transmittance of the ring resonator is the same as that of a Fabry-Perot interferometer.

The free spectrum range, FSR, which corresponds to the available frequency bandwidth, is given by:

$$\text{FSR} = \frac{c}{2\pi nr} \qquad (4.10)$$

For example, a ring radius r of 1 mm corresponds to an FSR of 25 GHz. The Q-value (finesse) of the resonator is:

$$Q = \frac{\pi\sqrt{R_r}}{1-R_r} \qquad (4.11)$$

Figure 4.2(c) shows an MZ filter with a ring resonator as in Oda, et al. (1988). The ring resonator is connected to the shorter arm of the MZ filter; the amplitude coupling ratio is k. The transmittance of this filter is described by:

$$T_{1-3} = \cos^2\left(\frac{\Phi-\theta}{2}\right) \qquad (4.12)$$

$$T_{1-4} = \sin^2\left(\frac{\Phi-\theta}{2}\right) \qquad (4.13)$$

where Φ is the phase delay of the ring resonator and θ is the round trip phase delay of the ring resonator. They are written as:

$$\Phi = \tan^{-1}\left\{\frac{(r^2-1)\sin\theta}{2r-(r^2+1)\cos\theta}\right\} \qquad (4.14)$$

$$\theta = \beta L_0 \qquad (4.15)$$

where $r = \sqrt{(1-k^2)}$, β is the propagation constant, and L_0 is the round trip time of the ring resonator. In addition, when ΔL and L_0 satisfy:

$$\beta \Delta L = \frac{\beta L_0}{2} \pm \frac{\pi}{2} \qquad (4.16)$$

and

$$\Delta L \frac{d\beta}{df} = \frac{d\Phi}{df} \qquad (4.17)$$

the critical coupling ratio $k = 2\sqrt{2}/3$ is obtained. In this case, the bandwidth (0.5 dB pass bandwidth) is expanded to as wide as 80% of the channel frequency separation, and the cut-off curve is sharpened compared to a simple MZ filter.

Figure 4.2(d) shows a double ring resonator described by Oda, Takato and Toba (1991). In this case, the relationship of the free spectrum range between the ring resonators is described as:

$$FSR = M_1 \cdot FSR_1 = M_2 \cdot FSR_2 \qquad (4.18)$$

where M_1 and M_2 are integers. As a result, the FSR of the ring resonator is expanded by means of the vernier effect.

As described before, it is possible for MZ filters to multi/demultiplex more than three frequencies by cascading these filters with different frequency separations as shown in Fig. 4.3. An N-channel frequency division multiplexed signal with a frequency spacing of Δf is divided into two groups with a frequency spacing of $2\Delta f$ by the first filter, while the second filters divide the signals into four groups with a frequency spacing of $4\Delta f$. In general, n-stage MZ filters multi/demultiplex $N = 2^n$ channels; 16 and 128 channels can be multi/demultiplexed by 4 and 7 stage MZ filters, respectively.

4.1.3 Tunable optical filters

Tunable optical filters can be applied to OFDM information distribution systems and frequency switching systems. Table 4.1 summarizes the performance of various optical filters, and the schematic of the filters is shown in Fig. 4.4. Waveguide-type multi-stage MZ filters, FP filters, acousto-optic (AO) filters, and laser diode (LD) filters are possible alternatives as tunable optical filters. The allowable number of channels is determined by the ratio of the tunable range to minimum channel spacing. MZ filters allow increased channel capacity by increasing the number of MZ filter stages. For example, one arbitrary signal can be selected from among 128 FDM signals with a seven-stage MZ filter as in

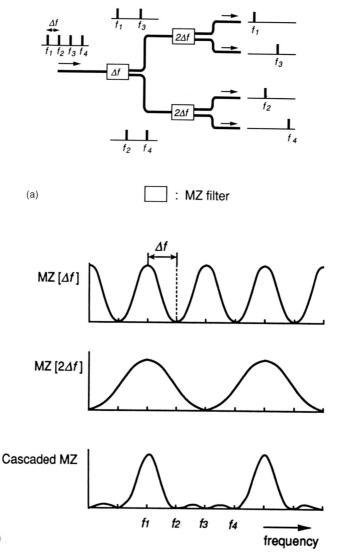

Fig. 4.3 Configuration of cascaded MZ filters.
(a) Source: Toba, *et al.*, 1986.

Oda, *et al.* (1990) and Takato, *et al.* (1990b). Frequency tuning is performed by applying power to the thin-film heaters, that are loaded on top of the waveguide arms, through the thermo-optic effect. The tuning speed is about 1 ms, and is limited by the thermal diffusion time from the electrode to the substrate.

A liquid crystal FP (LCFP) filter can be tuned to over 30 channels with applied voltages as low as 5 V according to Maeda, *et al.* (1990) and Hirabayashi, Tsuda

Table 4.1 Comparison of optical filters for optical FDM/WDM systems

Tunable filter	Tuning range (GHz)	Channel spacing (GHz)	Switching speed	Drive power	Reference
MZ	1280	10	1 ms	3.5 W	1
LCFP	7500	250	1 ms	<5 V	2, 3
FFP	15000	13.8	1 ms	10 V	4, 5
AOF	43000	200	10 ms	1 W	6
DFB-LD	100	10	1 ms	0.05 W	7, 8, 9

Sources:
1. Takato, et al. (1990)
2. Maeda, et al. (1990)
3. Hirabayashi, Tsuda and Kurokawa (1991)
4. Stone and Stulz (1987)
5. Eng, et al. (1990)
6. Smith, et al. (1990)
7. Numai, Murata and Mito (1988)
8. Vecchi, et al. (1988)
9. Nishio, et al. (1988)

and Kurokawa (1991). The tuning speed of the LCFP filter is of the order of milliseconds, and is limited by the time required to reorient the molecules. Nematic liquid crystals have birefringence due to the orientational order of the molecules. Thus polarization diversity or another polarization compensation scheme should be considered for use in practical optical FDM systems.

A fiber FP (FFP) filter consists of single-mode fiber and precision optical connector parts (Stone and Stulz, 1987). The fiber is cemented into three glass ferrules and the end polished flat. High reflecting dielectric mirrors are applied to the two faces indicated and antireflection coatings are applied to the other facet. The three ferrules are aligned by a precision split sleeve. A stacked piezoelectric transducer varies the mirror spacing L by increasing the gap; approximately 10 V scans through one FSR. The FFP filter is a compact device with a high resolution, finesse up to 200. Thus 30 channels can be tuned by a single FFP filter. A vernier configured FFP filter, similar to the double ring resonators described in section 4.1.1, can tune over 200 channels. Moreover, the FFP filter can be used for optical frequency discrimination as well as channel selection. Two-stage FFP filters consisting of a narrowband filter followed by a wideband filter have a tuning range of 15 000 GHz with a finesse of more than 5000 (Eng, et al., 1990). This filter is capable of covering over 1000 channels. The tuning time is 1 ms, and is limited by the response time of the piezo-electric element.

Acousto-optic (AO) filters use the collinear configuration to achieve TE-TM mode conversion and a wavelength resolution of about 1 nm (125 GHz frequency range at 1.55 μm wavelength range) and a tunable range of 300 nm (40 000 GHz frequency range) as in Smith (1990). That is, one channel from among more than 100 channels can be selected. The tuning speed of the AO filter is as fast as 1 μs. It has the unique multiwavelength filtering capability if multifrequency control signals are utilized.

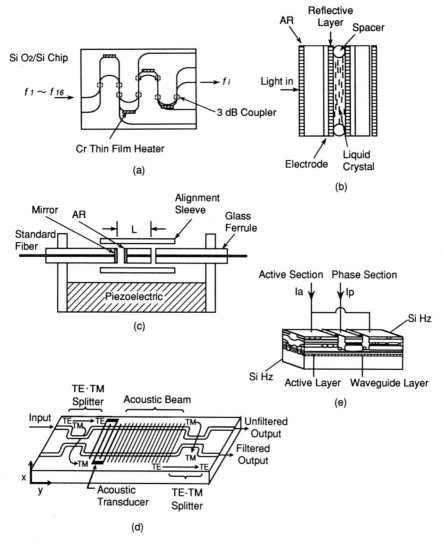

Fig. 4.4 Schematic of tunable filters: (a) serially connected MZ filters (for 16-channel signals); (b) liquid crystal FP filter; (c) fiber FP filter; (d) acousto-optic filter; (e) phase-shift-controlled DFB filter.
(Source: (a) Oda, et al., 1990; (b) Maeda, et al., 1990; (c) Stone, et al., 1987; (d) Smith, et al., 1990; (e) Numai, et al., 1988.)

A DFB LD filter with phase-shift control has a channel spacing as narrow as 10 GHz and a tuning speed as fast as a few nanoseconds (see Numai, Murata and Mito, 1988, and Vecchi, et al., 1988). Its gain allows the signal to be selected and amplified simultaneously. To date, an eight-channel filter has been reported by Nishio, et al. (1988). The channel capacity is limited by the continuous tuning range of the LD.

4.2 WAVEGUIDE-TYPE MACH-ZEHNDER INTERFEROMETER FILTERS

Masao Kawachi

Various types of optical filters for optical FDM or dense WDM have been proposed as listed in section 4.1. Recent developments in integrated silica waveguide technologies have utilized these filters in integrated-optic configurations (Kawachi, 1990). This section describes recent progress in silica waveguide filters with emphasis on Mach-Zehnder (MZ) interferometers which can handle optical frequency spacings of 10 GHz or less.

4.2.1 Basic interferometer structure

A typical configuration of a silica waveguide MZ interferometer filter on a silicon substrate is shown in Fig. 4.5. This interferometer filter consists of two directional couplers of two silica waveguide arms connecting the two couplers on a silicon substrate as given in Takato, et al. (1990). The optical frequency spacing $\Delta f(=f_1 - f_2)$ is related to the path difference ΔL between the two directional couplers by the following equation.

$$\Delta f = \frac{c}{2n\Delta L} \tag{4.19}$$

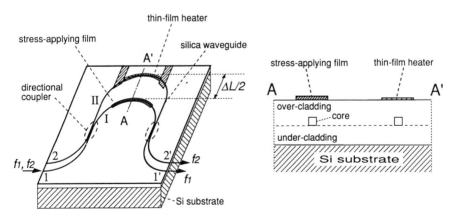

Fig. 4.5 Configuration of Mach-Zehnder silica waveguide interferometer filter showing cross-section A−A′.

where c is the light velocity and n the refractive index of the waveguide (see section 4.1). For example, $\Delta f = 10$ GHz for $n \sim 1.46$ and $\Delta L \sim 10$ mm.

The thermo-optic (i.e. temperature-dependent refractive index change) effect in silica glass realizes optical tuning or switching functions in a passive silica waveguide (Kawachi, 1990). For this purpose, one of the MZ interferometer arms is equipped with a thin-film Cr heater as illustrated in Fig. 4.5. When an electric voltage is applied to the thin-film heater, the refractive index of the heater waveguide increases and the optical path length changes by $(dn/dT)L\Delta T$. (dn/dT is the thermo-optic constant of the silica waveguide, L is the heater length and ΔT the temperature increase). For SiO_2, $dn/dT = 1 \times 10^{-5}$; for example, when a 10 mm long waveguide core is heated by ~ 7.8 °C, the optical path length changes by 0.78 μm and thus the phase shift is π for a 1.55 μm lightwave. The measured driving power for the π phase shift is typically 0.5 W on an Si substrate and the corresponding response time is 1–2 ms both for rise and fall. The Si substrate acts as the heat sink needed for stable thermo-optic operation.

Silica waveguides on an Si substrate exhibit a birefringence of about 4×10^{-4} between the TM and TE modes, because of the difference in the thermal expansion coefficient between doped SiO_2 ($0.5 - 1 \times 10^{-6}$/deg) and Si (2.5×10^{-6}/deg). To attain a polarization-insensitive MZ interferometer filter, the phase difference between the TM and TE modes along the two waveguide arms has to be adjusted to zero or $2N\pi$, and hence,

$$\int_I B dl_1 - \int_{II} B dl_2 = 2N\lambda \qquad (4.20)$$

where N is an integer and l_1 and l_2 represent line coordinates taken along the waveguide arms I and II. This adjustment can be achieved by depositing a stress-applying film (typically an RF-sputtered a-Si film) on part of the waveguide arm as illustrated in Fig. 4.5. This stress-applying film is trimmed by YAG laser beam irradiation so that the above equation is precisely satisfied (Sugita, et al., 1990).

This integrated-optic approach as illustrated in Fig. 4.5 is superior to conventional bulk-optic and fiber-optic approaches in terms of productivity, device stability and its suitability for multiple filter integration.

4.2.2 Silica waveguides

Various processes have been proposed for the fabrication of silica waveguides. Two major technologies are chemical vapor deposition (CVD) in Valette, et al. (1989) and Henry, Blonder and Kazarinov (1989) and flame hydrolysis deposition (FHD) in Kawachi (1990), both of which can form low-loss silica waveguides that match to optical fiber core dimensions.

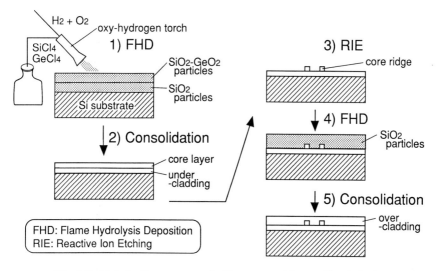

Fig. 4.6 Fabrication process of silica waveguides on Si substrate.

In the CVD method, doped silica glass films are deposited by the thermal oxidation of SiH_4 and PH_3 on silicon wafers heated in a reactor. The FHD method deposits fine glass particles, synthesized by flame hydrolysis of $SiCl_4$ and $GeCl_4$ in an oxy-hydrogen torch, onto silicon wafers as illustrated in Fig. 4.6. After deposition, the Si wafers with the porous glass layers are heated up to 1200–1300 °C in an electric furnace for consolidation. This method, originally developed for optical fiber fabrication, has the advantage of being able to make thick doped silica films (up to $\sim 100\,\mu m$ thick) with high deposition rates.

The waveguide core ridges are then defined by photolithography, corresponding to the MZ interferometer layout. Unnecessary portions of the deposited core glass layer are removed with reactive ion etching (RIE) or with reactive ion beam etching (RIBE). Finally, the core ridges are covered with an SiO_2 over-cladding layer deposited again either by CVD or FHD.

The silica waveguides thus fabricated on Si substrates have a core size of $5\,\mu m \times 5\,\mu m$ to $8\,\mu m \times 8\,\mu m$ which matches to single-mode optical fiber cores. The cores are completely embedded on a 40–60 µm thick SiO_2 cladding on an Si substrate. The propagation loss of the silica waveguides has reached a level of 0.01–0.1 dB/cm. The waveguide parameters for two kinds of SiO_2–GeO_2 buried waveguides (low Δ and high Δ waveguides) are listed in Table 4.2.

In constructing the desired MZ interferometers as shown in Fig. 4.5, bent waveguides with a small radius of curvature are required, and thus, the high Δ waveguides, which allow a minimum curvature radius of ~ 5 mm, are preferred for such applications.

116 OPTICAL FILTERS AND COUPLERS

Table 4.2 Fundamental characteristics of single-mode silica waveguides

	Low Δ	High Δ
Relative index difference Δ (%)	0.25	0.75
Core size (μm)	8 × 8	6 × 6
Loss (dB/cm)	<0.1	<0.1
Fiber coupling loss* (dB/point)	<0.1	~0.5
Minimum bending radius** (mm)	25	5

*Single-mode fiber used: 2a = 8.9 μm, Δ = 0.27%; with index-matching oil.
**Bending loss in a 90° arc waveguide is less than 0.1 dB at λ = 1.55 μm.

A cross-sectional view of two high Δ waveguides formed into the directional coupler element for an MZ interferometer is shown in Fig. 4.7. This well-defined structure, in which optical coupling takes place between the two adjacent cores a few micrometers apart, plays the important role of "half mirrors", splitting and combining light beams in the MZ interferometer. The 50% coupling length of the silica waveguide directional couplers, corresponding to a 1:1 optical power splitting ratio, is typically 0.5–1.0 mm.

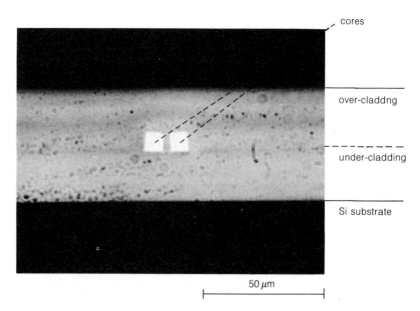

Fig. 4.7 Cross-sectional micrograph view of two adjacent silica waveguide cores as directional coupler in an MZ interferometer.

To create a practical device, the waveguide ends of the MZ interferometer filters have to be connected to input/output optical fibers with high efficiency and reliability. Silica waveguides of the same glass composition as the optical fibers appear to have an advantage over waveguides composed of other materials in terms of coupling efficiency.

4.2.3 Fundamental characteristics

The optical frequency response of an MZ filter fabricated in accordance with the above discussion is shown in Fig. 4.8; $\Delta L = 10$ mm and a 25 mm × 25 mm Si substrate is used. A tunable DFB laser diode (DFB-LD, $\lambda = 1.55\,\mu$m) was used as the light source. The optical frequency of the DFB-LD light was changed by sweeping the LD injection current, and the optical transmittance for path 1–1' in the MZ interferometer was monitored. Figure 4.8 confirms the $\Delta f = 10$ GHz demultiplexing function of the interferometer. The fiber-to-fiber total loss and the crosstalk attenuation of the interferometer filter were ~ 2.5 dB and ~ 20 dB respectively. The transmission characteristics were also evaluated by launching a fixed-frequency DFB-LD light and changing the applied electric power of the thin-film heater (the thermo-optic phase shifter).

Figure 4.9 shows the relation between the heater power and the transmittance loss for path 1–1'. The heater power required to change the optical output from maximum to minimum (or vice versa) was 0.52 W, corresponding to a π phase

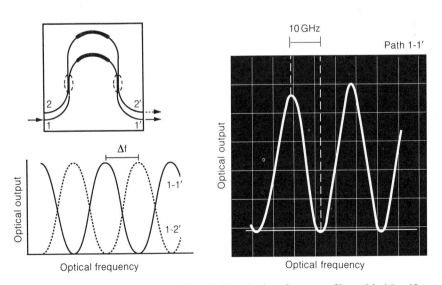

Fig. 4.8 Optical frequency response of Mach-Zehnder interferometer filter with $\Delta L = 10$ mm.

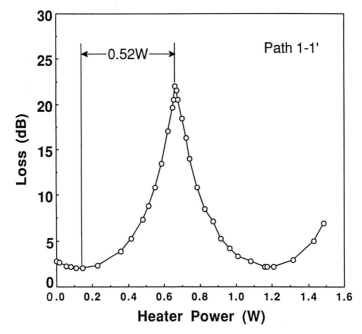

Fig. 4.9 Transmittance loss against heater power applied to thin-film heater for path 1 − 1′ in Mach-Zehnder interferometer filter.

shift. The tuning characteristics of the thin-film heater can allow the MZ filter to work as a frequency-selection switch.

MZ interferometers with larger path differences, $\Delta L = 20$ mm, 50 mm and 100 mm, have also been fabricated, corresponding to $\Delta f = 5$ GHz, 2 GHz and 1 GHz, respectively (Takato, et al., 1990). Their Si substrate sizes are 25 mm × 30 mm, 25 mm × 45 mm and 50 mm × 50 mm, respectively. Filters with such large ΔL values, and consequently small Δf, find use also as optical frequency discriminators in optical FSK transmission systems.

4.2.4 Integration of multiple interferometer units

In the integrated waveguide approach as described above, it is easy to integrate a number of MZ interferometer filters on a single Si substrate. Seven interferometer units with seven different ΔL values of 10 mm, 5 mm, 2.5 mm, 1.25 mm, 0.63 mm, 0.31 mm and 0.16 mm were, for example, connected in series to form a 10 GHz spaced 128-channel frequency-selection filter by Takato, et al. (1990), as schematically shown in Fig. 4.10. The waveguide layout was designed to be 50 mm × 60 mm in size.

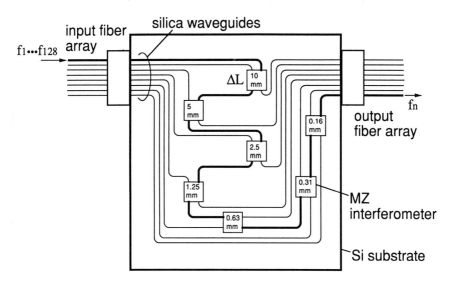

Fig. 4.10 Configuration of 128-channel frequency selection filter integrated onto one silicon substrate.

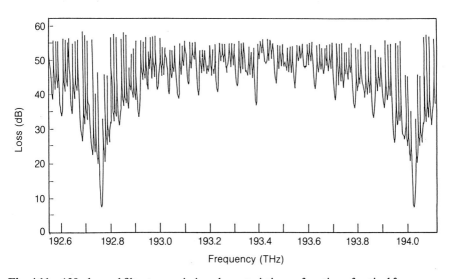

Fig. 4.11 128-channel filter transmission characteristic as a function of optical frequency.

Figure 4.11 shows the filter transmission characteristics for non-polarized light as a function of optical frequency. They were measured by changing the frequency of ten DFB laser diodes by temperature manipulation. The free spectral range was 1280 GHz. Interchannel crosstalk levels between the transmitted frequency and frequencies with 10 GHz intervals ranged from −20 to less than −50 dB. The fiber-to-fiber insertion loss was 6.7 dB. The total crosstalk level is estimated to be less than −13 dB for the selected channel when 128 optical carriers with a channel spacing of 10 GHz are launched to the filter.

Another useful configuration is a 2^N channel multiplexer consisting of $1, 2, 4, \cdots$, and 2^{N-1} MZ filter units with a path difference of ΔL, $\Delta L/2$, $\Delta L/4, \cdots$, and $\Delta L/2^{N-1}$, respectively. A 10 GHz spaced eight-channel multiplexer, which consisted of one, two and four MZ units with a path difference of 10 mm, 5 mm and 2.5 mm respectively, was successfully fabricated with fiber-to-fiber losses of 4.7 to 5.8 dB according to Takato (1990).

It seems worthwhile to note finally that the present integrated silica waveguide technologies have been successfully applied also to the fabrication of optical ring resonators, transversal filters and arrayed waveguide multi/demultiplexers (see section 4.1), though we would not have enough space to cover them in this section. Integrated MZ interferometer filters and related devices are expected to play important roles in constructing efficient optical FDM networks.

4.3 OPTICAL COUPLERS

Masaru Kawachi

Various kinds of optical couplers are required in constructing coherent optical communication systems. These couplers include 2×2 couplers with various coupling ratios, $1 \times N$ power splitters for distributing optical signals and $N \times N$ star couplers for mixing and distributing multiple-channel optical signals.

There are three major approaches, as illustrated in Fig. 4.12, for making optical couplers. These are (a) the bulk-optic approach which uses prisms and lenses, (b) the fiber-optic approach which mainly employs fused fiber couplers, and (c) the integrated-optic approach which constructs fiber-matched waveguides on planar substrates. To date, the fiber-optic approach has received the most commercial attention for producing low-loss 2×2 couplers and their derivatives. Integrated silica waveguide technologies, on the other hand, have only recently reached a level of development sufficient to produce practical waveguide-type

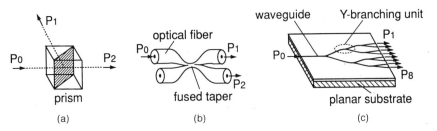

Fig. 4.12 Three approaches to making optical couplers: (a) bulk-optic; (b) fiber-optic; (c) integrated-optic.

couplers such as 1 × 8 and 8 × 8 that can compete with conventional bulk-optic and fiber-optic couplers. The following section describes the status of fiber-optic couplers and integrated-optic waveguide-type couplers.

4.3.1 Fiber-optic couplers

Figure 4.13 shows two different technologies for fabricating 2 × 2 couplers found in Kawachi (1985/1986). In the polishing method, two single-mode fibers are polished to within a few microns from the core center line. The two polished surfaces are then placed in contact with each other as shown in Fig. 4.13(a). The strength of evanescent light coupling can be varied over the range of $0 \sim 100\%$ by moving the fibers out of or into alignment with each other.

In the fusion/tapering method, two adjacent single-mode fibers are partly fused together with a micro-torch and then elongated so that the fused region has a biconical taper structure as shown in Fig. 4.13(b). Continued monitoring of the optical signals from output fibers during the fusion/tapering process enables the process to be stopped when the desired coupling ratio is achieved between them. The taper waist diameter is typically $10 \sim 20\,\mu m$. Packaging results in a component that is rugged and environmentally stable.

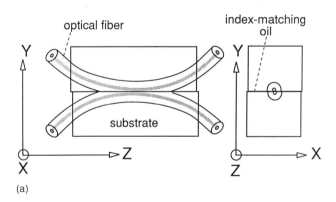

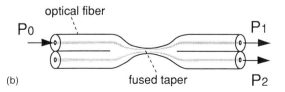

Fig. 4.13 Two different techniques for making fiber-optic 2 × 2 couplers: (a) polishing method; (b) fusion/tapering method.

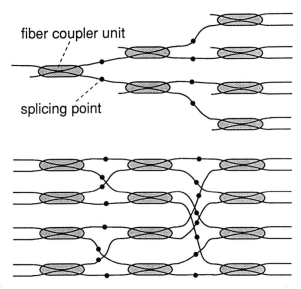

Fig. 4.14 Fiber-optic array configurations of (a) 1 × 8 splitter; (b) 8 × 8 coupler.

The commercial 2 × 2 couplers thus fabricated to date with the above methods have excess losses lower than 0.5 dB. Excess losses even as low as 0.1 dB have been achieved with the fusion/tapering method. The most popular coupler has a 50% coupling ratio corresponding to a 1:1 power splitting ratio, and is called a 3 dB coupler.

Multiple 2 × 2 fiber couplers (3 dB couplers) can be combined together to form larger scale couplers as exemplified in Fig. 4.14, where (a) seven 3 dB couplers are cascaded to form a 1 × 8 splitter and (b) twelve 3 dB couplers form an 8 × 8 star coupler. In this manner, a series of optical splitters (1 × 4, 1 × 8, 1 × 16, 1 × 32, 1 × 64 and 1 × 128) and star couplers (4 × 4, 8 × 8, 16 × 16, 32 × 32, 64 × 64 and 128 × 128) have been constructed.

One common problem in constructing large-scale splitters and star couplers in fiber-optic configurations is that troublesome fiber handling and splicing procedures are required for processing large numbers of 3 dB coupler units. The number of 3 dB couplers needed in an N × N star coupler is $(N/2)\log_2 N$, i.e. 448 3 dB couplers are necessary to construct a 128 × 128 star coupler.

4.3.2 Integrated-optic waveguide couplers

One possible solution that overcomes the problem encountered in constructing large-scale fiber-optic couplers is to use integrated waveguide technologies to form waveguide-type couplers as shown in Fig. 4.12(c).

OPTICAL COUPLERS

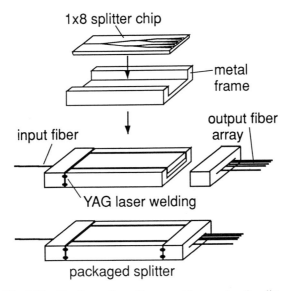

Fig. 4.15 Configuration of waveguide-type 1 × 8 splitter.

Figure 4.15 shows the configuration of a waveguide-type 1 × 8 splitter. Seven units of Y branching silica waveguides are integrated together on a single Si substrate to form a 1 × 8 optical power splitter. The waveguide pitch (the center-to-center distance of two adjacent waveguides) at the substrate end is 250 μm, compatible with that of single-mode optical fiber arrays. The fiber array frames are fixed to the waveguide ends by using a YAG laser welding technique as shown in Fig. 4.15. The excess loss for a packaged 1 × 8 splitter, fabricated with low Δ silica waveguides (Table 4.2) on a 5 mm × 26 mm Si substrate, was ~1 dB ± 0.5 dB including input/output optical fiber coupling losses (Kobayashi, et al., 1990).

Figure 4.16 shows (a) the waveguide layout and (b) the beam propagation method simulation (BPM: a numerical method to simulate the lightwave propagation in axially varying waveguides such as tapers, bends and branches) of an 8 × 8 coupler consisting of two fan-shaped channel waveguide arrays facing each other with a slab waveguide region joining them. The waveguide interval gradually decreases in the array region and becomes a few micrometers at the interface of the array guides and the slab region. Because the radiation pattern is the Fraunhofer pattern (Fourier transform) of the field profile at the input side slab-array interface, proper sidelobes must be produced by the mode coupling from the excited input waveguide to neighboring guides. The peripheral waveguides need dummy guides to guarantee the same coupling condition as the central guides. The star coupler parameters such as the aperture angle and the separation between the two arrays

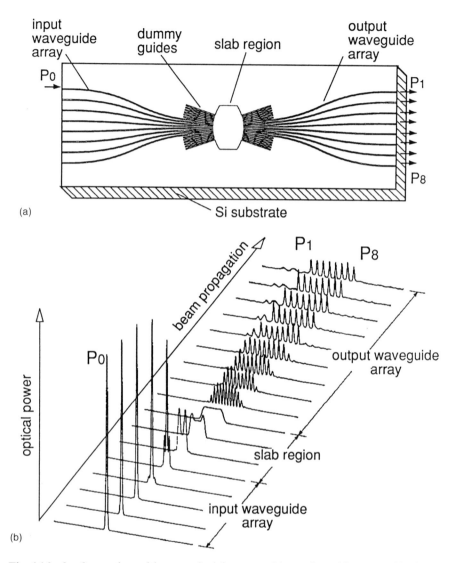

Fig. 4.16 8 × 8 coupler with central slab waveguide region: (a) waveguide layout; (b) BPM simulation of waveform transients when light is coupled into peripheral waveguide.

have been optimized so as to achieve maximum output and good coupling uniformity.

The 8 × 8 coupler, fabricated on a 5 mm × 26 mm Si substrate, exhibits low loss (1.42 dB average excess loss in addition to 8 dB intrinsic splitting loss) and good coupling uniformity (standard deviation: 0.49 dB), which agree quite well with the BPM simulation in Okamoto, *et al.* (1991). This performance is comparable to conventional fiber-type 8 × 8 couplers in which 12 units of 2 × 2 fused fiber couplers are combined together and packaged using somewhat troublesome fiber splicing procedures.

The present planar-type compact coupler geometry has the potential to produce very large-scale one-chip N × N star couplers as reported by Dragone, *et al.* (1989). The advantage of the present integrated-optic approach is much more obvious when N becomes as large as 16, 32, 64 or 128.

REFERENCES

Dragone, C., Edwards, C. A. and Kistler, R. C. (1991) Integrated optics N × N multiplexer on silicon, *IEEE Photonics Technology Letters*, **3**, pp. 896–899.

Dragone, C., Henry, C. H., Kaminow, I. P. and Kistler, R. C. (1989) Efficient multichannel integrated optic star coupler on silicon, *IEEE Photonics Technology Letters*, **1**, pp. 241–243.

Eng, K. Y., Santoro, M. A., Koch, T. L., Stone, J. and Snell, W. W. (1990) Star-coupler-based optical cross-connect switch experiments with tunable receivers, *IEEE Journal Selected Areas Communications*, **8**, pp. 1026–1031.

Flanders, D. C., Kogelnik, H., Schmidt, R. V. and Shank, C. V. (1974) Grating filters for thin-film optical waveguides, *Applied Physics Letters*, **24**, pp. 194–196.

Frenkel, A. and Lin, C. (1989) Angle-tuned etalon filters for optical channel selection in high density wavelength division multiplexed systems, *IEEE Journal Lightwave Technology*, **7**, pp. 615–624.

Henry, C. H., Blonder, G. E. and Kazarinov, R. F. (1989) Glass waveguides on silicon for hybrid optical packaging, *IEEE Journal Lightwave Technology*, **7**, pp. 1530–1539.

Henry, C. H., Shani, Y., Kistler, R. C., Jewell, T. E., Pol, V., Olsson, N. A., Kazarinov, R. F. and Orlowsky, K. J. (1989) Compound Bragg reflection filters made by spatial frequency doubling lithography, *IEEE Journal Lightwave Technology*, **7**, pp. 1379–1385.

Hirabayashi, K., Tsuda, H. and Kurokawa, T. (1991) Narrow-band tunable wavelength-selective filters of Fabry-Perot interferometer with a liquid crystal intracavity, *IEEE Photonics Technology Letters*, **3**, pp. 213–215.

Inoue, K., Takato, N., Toba, H. and Kawachi, M. (1988) A four-channel optical waveguide multi/demultiplexer for 5 GHz-spaced optical FDM transmission, *IEEE Journal Lightwave Technology*, **6**, pp. 339–345.

Ishio, H., Minowa, J. and Nosu, K. (1984) Review and status of wavelength-division-multiplexing technology and its application, *IEEE Journal Lightwave Technology*, **LT-2**, pp. 448–463.

Kanada, T., Okano, Y., Aoyama, K. and Kitami, T. (1983) Design and performance of WDM transmission systems at 6.3 Mb/s, *IEEE Transactions on Communications*, **COM-31**, pp. 1095–1102.

Kawachi, M. (1985/1986) Fiber-optic components, *Optical Devices & Fibers*, **JARECT 17**, OHM North-Holland, pp. 171–180.

Kawachi, M. (1990) Silica waveguides on silicon and their application to integrated-optic components, *Optical and Quantum Electronics*, **22**, pp. 49–52.

Kobayashi, S., Kito, T., Hida, Y. and Yamaguchi, M. (1990) Optical waveguide 1 × 8 splitter module, *NTT R&D*, **39**, pp. 931–938 (in Japanese).

Maeda, M. W., Patel, J. S., Lin, C., Horrobin, J. and Spicer, P. (1990) Electronically tunable liquid-crystal-etalon filter for high-density WDM systems, *IEEE Photonics Technology Letters*, **2**, pp. 820–822.

Minowa, J. and Fujii, Y. (1983) Dielectric multilayer thin-film filters for WDM transmission systems, *IEEE Journal Lightwave Technology*, **LT-1**, pp. 116–121.

Nishio, M., Numai, T., Suzuki, S., Fujiwara, M., Itoh, M. and Murata, S. (1988) Eight-channel wavelength-division-switching experiment using wide-tuning-range DFB LD filters, in *Proceedings ECOC'88*, part 2, pp. 49–52.

Nosu, K., Toba, H. and Iwashita, K. (1987) Optical FDM transmission technique, *IEEE Journal Lightwave Technology*, **LT-5**, pp. 1301–1308.

Numai, T., Murata, S. and Mito, I. (1988) 1.5 µm tunable wavelength filter using a phase-shift-controlled distributed feedback laser diode with a wide tuning range and a high constant gain, *Applied Physics Letters*, **54**, pp. 1859–1860.

Oda, K., Takato, N., Kominato, T. and Toba, H. (1990) A 16-channel frequency selection switch for optical FDM distribution systems, *IEEE Journal Selected Areas Communications*, **8**, pp. 1132–1140.

Oda, K., Takato, N. and Toba, H. (1991) A wide-FSR waveguide double-ring resonator for optical FDM transmission systems, *IEEE Journal Lightwave Technology*, **9**, pp. 728–736.

Oda, K., Takato, N., Toba, H. and Nosu, K. (1988) A wide-band guided-wave periodic multi/demultiplexer with a ring resonator for optical FDM transmission systems, *IEEE Journal Lightwave Technology*, **6**, pp. 1016–1023.

Ohmori, Y., Kominato, T., Okazaki, H. and Yasu, M. (1990) Low loss GeO_2-doped silica waveguides for large scale integrated optical devices, in *Technology Digest, OFC'90*, San Francisco, CA, paper W2.

Ohtomo, I., Shimada, S. and Suzuki, N. (1971) Two-cavity ring-type channel-dropping filters for a millimeter-wave guided wave communication system, *IEEE Transactions Microwave Theory Technology*, **MTT-19**, pp. 481–484.

Okamoto, K., Takahashi, H., Suzuki, S. and Ohmori, Y. (1991) Design and fabrication of integrated-optic 8 × 8 star coupler, *Electronics Letters*, **27**, pp. 774–775.

Sasayama, K., Okuno, M. and Habara, K. (1989) Coherent optical transversal filter using silica-based single-mode waveguides, *Electronics Letters*, **22**, pp. 1508–1509.

Sasayama, K., Okuno, M. and Habara, K. (1992) Frequency-division-multiplexing multichannel selector using a coherent optical transversal filter, *OFC'92 Technology Digest*, **TuC2**, San Jose, CA, pp. 10–11.

Smith, D. A., Baran, J. E., Johnson, J. J. and Cheung, K. W. (1990) Integrated-optic acoustically-tunable filters for WDM networks, *IEEE Journal Selected Areas Communications*, **8**, pp. 1151–1159.

Stone, J. and Stulz, L. W. (1987) Pigtailed high-finesse tunable Fabry-Perot interferometers with large, medium and small free spectral ranges, *Electronics Letters*, **23**, pp. 781–783.

Sugita, A., Jinguji, K., Takato, N. and Kawachi, M. (1990) Laser-trimming adjustment of waveguide birefringence in optical FDM components, *IEEE Journal Selected Areas in Communications*, **8**, pp. 1128–1131.

Takahashi, H., Nisi, I. and Hibino, Y. (1992) 10 GHz spacing optical frequency division multiplexer based on arrayed-waveguiding grating, *Electronics Letters*, **28**, pp. 380–382.

Takahashi, H., Suzuki, S., Kato, K. and Nishi, I. (1990) Arrayed-waveguide grating for wavelength division multi/demultiplexer with nanometre resolution, *Electronics Letters*, **26**, pp. 87–88.

Takahashi, H., Suzuki, S. and Nishi, I. (1991) Multi/demultiplexer for nanometer-spacing WDM using arrayed-waveguide grating, *Technology Digest, IPR'91*, Monterey, CA, post deadline paper PD-1.

Takato, K., Jinguji, K., Yasu, M., Toba, H. and Kawachi, M. (1988) Silica-based single-mode waveguides on silicon and their application to guided-wave optical interferometers, *IEEE Journal Lightwave Technology*, **6**, pp. 1003–1010.

Takato, N., Kominato, T., Sugita, A., Jinguji, K., Toba, H. and Kawachi, M. (1990a) Silica-based integrated optic Mach-Zehnder multi/demultiplexer family channel spacing of 0.01–250 nm, *IEEE Journal Selected Areas Communications*, **8**, pp. 1120–1127.

Takato, N., Sugita, A., Onose, K., Okazaki, H., Okuno, M., Kawachi, M. and Oda, K. (1990b) 128-channel polarization-insensitive frequency-selection-switch using high-silica waveguides on Si, *IEEE Photonics Technology Letters*, **2**, pp. 441–443.

Toba, H., Oda, K. and Nosu, K. (1988) 5 GHz-spaced, eight-channel optical FDM transmission experiment using guided-wave tunable demultiplexer, *Electronics Letters*, **24**, pp. 78–80.

Valette, S., Renard, S., Denis, H. Jadot, J. P., Founier, A., Philippe, P., Gidon, P., Grouillet, A. M. and Desgranges, E. (1989) Si-based integrated optics technologies, *Solid State Technology*, Feb., pp. 69–74.

Vecchi, M. P., Kobrinski, H., Goldstein, E. L. and Bulley, R. M. (1988) Wavelength selection with nanosecond switching times using distributed-feedback optical amplifiers, in *Proceedings ECOC'88*, Brighton, England, pp. 247–250.

Wisely, D. W. (1991) 32-channel WDM multiplexer with 1 nm channel spacng and 0.7 nm bandwidth, *Electronics Letters*, **27**, pp. 520–521.

5
Optical frequency division multiplexing systems

5.1 OPTICAL FREQUENCY STABILIZATION AND MEASUREMENT

Osamu Ishida

In frequency division multiplexing (FDM) communication schemes, carrier frequency stability is essential since the frequency values themselves contain the information needed for channel selection. Higher transmission capacity per unit frequency requires higher carrier frequency stability.

In the lightwave region, a single-mode laser must be used as the carrier frequency generator. The laser frequency v [Hz] is related to the wavelength λ [m] by:

$$c = v\lambda \tag{5.1}$$

where c is the speed of light. Note that c in vacuum is a defined constant (2.99792458×10^8 m/s) since 1983. Frequencies are as high as 187–200 THz (1 THz = 10^{12} Hz) in the wavelength region of 1.50–1.60 μm where silica fiber exhibits the lowest transmission losses (< 0.2 dB/km); 1.00 nm in this wavelength region corresponds to 117–133 GHz. These frequencies are more than 10^3 times higher than conventional microwave frequencies.

We restrict our interest to the 1.5 μm diode lasers suitable for FDM applications. This subsection is organized as follows: laser frequency fluctuations and stabilization techniques are described in 5.1.1 and 5.1.2 respectively, while laser frequency measurement techniques are reviewed in 5.1.3.

5.1.1 Laser frequency fluctuations

Before discussing laser diode fluctuations, we need to characterize frequency fluctuations. Assume that the instantaneous frequency v fluctuates around the nominal value v_0 as

$$v(t) = v_0 + v_n(t) \qquad (5.2)$$

where the stochastic fluctuation term $v_n(t)$ is assumed to have a zero mean by systematically excluding the drift term. Two measures are often used for analyzing $v_n(t)$: the power spectral density $S_v(f)$ and the two-sample variance $\sigma_v^2(t)$ as in Barnes, et al. (1971) and Kartaschoff (1978).

The power spectral density $S_v(f)$ [Hz²/Hz], where f is the Fourier frequency, is derived by the Fourier transform of the autocorrelation function of $v_n(t)$. This is the frequency domain measure of frequency fluctuation. A simple power series model:

$$S_v(f) = k_w + k_f f^{-1} + k_r f^{-2} \quad \text{(one-sided)} \ [\text{Hz}^2/\text{Hz}] \qquad (5.3)$$

often well fits practical oscillators according to Kartaschoff (1978). The terms proportional to f^0, f^{-1}, and f^{-2} are called white, flicker, and random walk fluctuations, respectively. Note that pure white fluctuations yield Lorentzian line shapes with a full width at half maximum (FWHM) of $\Delta v = \pi k_w$.

The two-sample variance $\sigma_v^2(\tau)$ [Hz²], on the other hand, is a time domain measure estimated from n-times successive frequency measurements as:

$$\sigma_v^2(\tau) = \lim_{n \to \infty} \frac{1}{2(n-1)} \sum_{i=1}^{n-1} (v_{i+1} - v_i)^2 \qquad (5.4)$$

where v_i is the averaged $v_n(t)$ over the i-th time interval of τ. Its normalized version $\sigma_y^2(\tau) = \sigma_v^2(\tau)/v_0^2$ is the well-known Allan variance.

The two measures $S_v(f)$ and $\sigma_v^2(\tau)$ are related as indicated by Barnes, et al. (1971) as:

$$\sigma_v^2(\tau) = 2 \int_0^\infty S_v(f) \frac{\sin^4(\pi f \tau)}{(\pi f \tau)^2} df \qquad (5.5)$$

Note that $S_v(f)$ cannot be derived from $\sigma_v^2(\tau)$ in general. Providing that $S_v(f)$ is simply modeled as equation 5.3, however, $S_v(f)$ can be estimated from $\sigma_v^2(\tau)$. Figure 5.1 shows the relations schematically. The value $\sigma_v^2(\tau)$ is easier to measure than $S_v(f)$ for slow fluctuations.

Now let us consider the frequency fluctuations of single-mode diode lasers. The diode laser frequency fluctuates through refractive index changes in the laser cavity. The refractive index n of the semiconductor laser cavity depends on both the carrier density and temperature. In single-electrode DFB lasers, the temperature fluctuations caused by current injection or device heating/cooling usually produce fluctuations slower than 1 kHz: typical coefficients are -1 GHz/mA and -10 GHz/K, respectively.

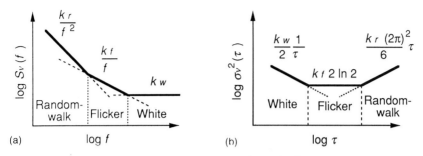

Fig. 5.1 Frequency fluctuation representations (schematic diagrams): (a) power spectral density $S_\nu(f)$ [Hz^2/Hz]; (b) two-sample variance $\sigma_\nu^2(\tau)$ [Hz^2] of the frequency averaged over time interval τ.

The upper trace in Fig. 5.2(a) shows typical DFB laser frequency fluctuations $\nu_n(t)$. The laser was packaged with hermetic sealing, and its temperature was held constant to within ± 0.002 K. Peak-to-peak fluctuations of more than 50 MHz were observed. Their two-sample variance values are plotted in Fig. 5.2(b) (open circles); flicker fluctuations predominate.

5.1.2 Laser frequency stabilization

(a) Extended-cavity laser

Diode laser frequency fluctuations can be suppressed if the refractive index fluctuations of the laser cavity are made less effective. This is achieved by using an external diffractive grating that extends the laser cavity length as in Wyatt and Delvin (1983). The inset in Fig. 5.3(a) shows the schematic of an extended cavity diode laser. Figures 5.3(a) and (b) show the residual frequency fluctuation $\nu_n(t)$ and root two-sample variance $\sigma_\nu(\tau)$ respectively. The drift observed in Fig. 5.3(a) is attributed to the piezoelectric transducer used for external grating control.

This technique also enables laser linewidth narrowing (i.e. white frequency noise reduction) and wide-range frequency tuning. However, the laser loses its compactness and becomes much more sensitive to mechanical vibration according to Al-Chalabi, *et al.* (1990).

(b) Feedback stabilization

Another approach is feedback stabilization to an external reference. Figure 5.4(a) shows the basic diagram: the laser frequency is discriminated by using an external reference, and the detected error signal is fed back to the diode laser.

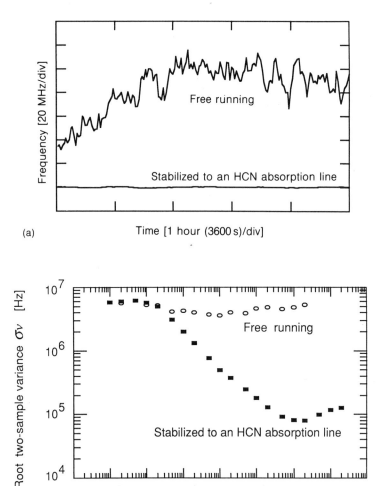

Fig. 5.2 Examples of laser diode frequency fluctuations and feedback stabilization (a) $v_n(t)$; (b) square-root of two-sample variance $\sigma_v^2(\tau)$.

There are several combinations of the (a) external reference, (b) error-signal generation, and (c) feedback technique. The best absolute stability (estimated in $\sigma_v(\tau)$) ever reported in the 1.5 μm region is of the order of 10^5 Hz (10^{-9} relative to the optical frequency) according to Ishida and Toba (1991a and 1991b).

External references. Atomic-transition or molecular absorption lines offer excellent long-term stability and are discussed hereafter. Periodic resonance character-

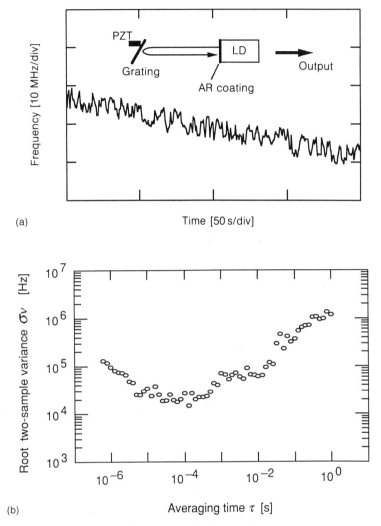

Fig. 5.3 Reduced fluctuations in an extended-cavity laser (a) $v_n(t)$; (b) square-root of two-sample variance $\sigma_v^2(\tau)$.

istics of a passive Fabry-Perot or ring resonator will be discussed in section 5.2. Such resonators are suitable for improving short-term stability or for equi-spaced multi-carrier stabilization.

Figure 5.4(b) shows how to detect molecular absorption lines: the gas cell is filled with molecular gas at low pressure (< 100 Torr). In the 1.5 μm region, acetylene (C_2H_2) as in Baldacci, Ghersetti and Rao (1977) and hydrogen cyanide (HCN) as in Sasada and Yamada (1990) show many absorption lines that

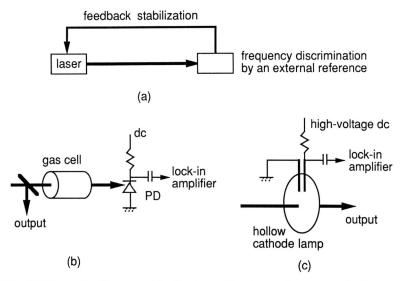

Fig. 5.4 (a) Schematic diagram of feedback stabilization against external references; (b) molecular reference; (c) atomic reference.

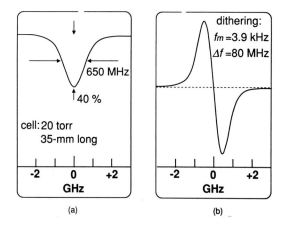

Fig. 5.5 C_2H_2 absorption lines: (a) C_2H_2 absorption line at 1535.39 nm; (b) lock-in-detected first derivative of (a).

originate mainly from the vibration of their ≡C–H bond. The absorption lines exist every 50–90 GHz, and are shifted if isotopic molecules ($^{13}C_2H_2$, $^{13}H^{15}CN$ etc.) are used. Figure 5.5(a) shows the C_2H_2 line centered at 1535.39 nm. The frequency stability of these references is restricted by the pressure and temperature shifts; reported values are 0.2 MHz/Torr and 0.1 MHz/K for C_2H_2 in Sakai, Kano and Sudo (1990).

Atomic transition lines can be employed through their optogalvanic spectrum. Figure 5.4(c) shows the schematic setup for optogalvanic signal detection; a hollow cathode lamp is filled with krypton (Kr) gas, and detects the electron transition in Kr atoms through a change in impedance (Chung, 1990). The simple configuration is its main advantage because few transitions exist in the 1.5–1.6 µm region.

Error-signal generation. Laser frequency dithering followed by synchronous (lock-in) detection is popular for obtaining zero cross-discrimination characteristics. Figure 5.5(b) shows the lock-in-detected first derivative of the C_2H_2 absorption line shown in Fig. 5.5(a). This technique offers error signals with high signal-to-noise ratios (S/N), while the short-term laser frequency stability is deteriorated by the laser frequency dithering.

Dithering the reference characteristics themselves, if possible, avoids laser dithering. The Kr optogalvanic spectrum can be dithered by magnetic fields through the Zeeman effect according to Chung and Derosier (1990). External optical frequency modulation is another approach applicable even for molecular absorption lines (Yanagawa, *et al.*, 1985).

The absorption slope itself is available for frequency discrimination so that laser dithering might not be needed. The resultant stability, however, is restricted by the large baseband fluctuations originating from optical power fluctuations and/or the detector's DC drift. Long-term stability better than 10^6 Hz is thus difficult, if not impossible, to achieve.

Feedback technique. Feedback frequency tuning of diode lasers is performed by changing the refractive index *n* through thermal and/or carrier effects as in Ohtsu (1988). Current injection causes the above two effects simultaneously, whereas a thermal cooler/heater and incoherent lightwave injection individually cause the thermal and carrier effects, respectively.

Thermal tuning through current injection is popular since electrical error-signal processing is easy to employ. Thermal tuning restricts the feedback bandwidth to less than 10 kHz. Faster feedback stabilization requires sophisticated laser diodes such as multisection DFB lasers and DBR lasers; current tuning in these lasers is dominated not by the thermal effect but by the carrier effect, and hence the response is uniform up to 100 MHz as in Kobayashi and Mito (1988). Incoherent lightwave injection is another approach for carrier-effect tuning; this can be employed even with conventional DFB lasers (Yasaka and Kawaguchi, 1988, and Inoue, 1990).

(c) Lightwave frequency synthesizer

The next stage of frequency stabilization is to construct a frequency synthesizer system that offers a frequency-tunable output. A two-laser configuration has been proposed for this purpose by Ishida and Toba (1991). Tunable laser frequency

136   OPTICAL FREQUENCY DIVISION MULTIPLEXING SYSTEMS

fluctuations are transferred to an offset-locked tracking laser as in Ishida (1991) and detected against molecular absorption lines. The residual fluctuations are shown in Fig. 5.2. Frequency tuning is performed by using offset frequency values that would be set to about ±50 GHz. Since molecular absorption lines exist every 50–90 GHz, almost all frequency values could be covered with this system.

5.1.3 Laser frequency measurement

Remember the basic concept of measurement: the unknown quantity is compared against a reference which has been calibrated with a standard. The caesium (Cs) atomic clock (9.192631770 GHz) has been the frequency standard since 1967. Moreover, since 1983, the Cs clock is also the wavelength standard as derived by equation 5.1 as in *Metrologia* (1984).

The optical frequency or wavelength references, therefore, eventually require calibration against a Cs clock. These calibrations, which are often called absolute optical frequency measurements, have been successfully performed using the "frequency chain" for several gas or dye lasers as in Jennings, Evenson and Knight (1986).

Table 5.1 shows typical results. In the 1.5 μm region, however, secondary standard lasers calibrated by the absolute measurement have not yet been authorized. Two promising approaches are:

1. frequency chain (Kourogi, *et al.*, 1991);
2. frequency division (Telle, Meschede and Hänsch, 1990).

Optical frequency conversion by nonlinear devices is the key technology for these approaches.

Let us switch our attention to a practical instrument for frequency or wavelength measurement in the 1.5 μm region. Table 5.2 summarizes three popular measuring instruments for laser diodes; these instruments have been calibrated by interferometric techniques, not by absolute measurement, and hence their accuracies are far below that of the Cs frequency standard.

Table 5.1 Optical frequency calibration readings

λ (nm)	Laser	Stabilization reference				ν (THz)
3392	He–Ne	CH_4	ν_3	P(7)	$F_2^{(2)}$	88.376 181 608
633	He–Ne	$^{127}I_2$	11–5	R(127)	i	473.612 214 8
612	He–Ne	$^{127}I_2$	9–2	R(47)	o	489.880 355 1
576	Dye	$^{127}I_2$	17–1	P(62)	o	520.206 808 51
525	Ar^+	$^{127}I_2$	43–0	P(13)	a_3	582.490 603 6

Table 5.2 Popular measuring instruments for laser diodes

Device	Reference	Comparison technique	Resolution	Range
Monochrometer	Grating	Mechanical	>2 GHz (10^{-5})	>100 THz
Wavelength counter	Stabilized laser	Mechanical and electrical	>200 kHz (10^{-9})	>100 THz
Frequency counter	Stabilized laser	Electrical	<200 kHz (10^{-9})	<0.1 THz

(a) Monochrometer (spectrometer or optical spectrum analyzer)

A diffraction grating is the reference; the input lightwave is diffracted by the grating and its diffraction angle is taken to indicate the wavelength and hence its frequency. The resolution and accuracy are restricted to the order of 2 and 20 GHz (10^{-5} and 10^{-4} relative to the optical frequency) respectively. The grating's diffractive selectivity limits the resolution, and the mechanical stability of the grating stage limits the accuracy. The possible measurement range is wider than 100 THz.

(b) Wavelength counter

Figure 5.6 shows a schematic diagram of the wavelength counter. The input lightwave wavelength is compared against the reference lightwave by using a moving Michelson interferometer discussed in Hall and Lee (1976). Movement of the corner cube causes the fringe numbers to change in proportion to the wavelengths. The input wavelength is, therefore, estimated by digitally comparing the fringe number changes. An He–Ne gas laser at 0.633 µm can be employed for the reference lightwave whose stability can be improved up to 0.04 MHz (10^{-10} relative to the optical frequency). The mechanical imperfection of the moving interferometer restricts its ultimate wavelength resolution and accuracy to the order of 0.2 MHz (10^{-9} relative to the optical frequency) (Ishikawa, *et al.* (1986)).

(c) Frequency counter

The most accurate comparison against a reference laser is performed by counting the optically-heterodyned beat note. The resolution is dominated by the reference laser stability itself since electric beat note counters offer sub-kilohertz resolution and accuracy.

The measurable frequency range is, however, restricted by the photodetector (PD) bandwidth for beat note detection. The practical PD bandwidth is as small as 50 GHz in the present state-of-the-art devices, and hence requires different

138 OPTICAL FREQUENCY DIVISION MULTIPLEXING SYSTEMS

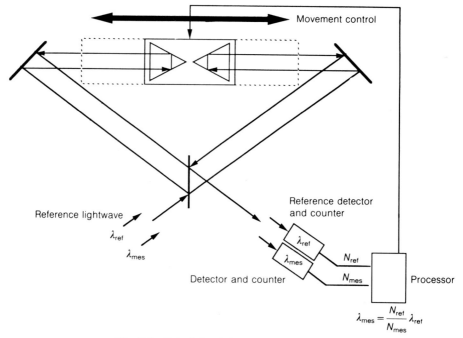

Fig. 5.6 Principle of the wavelength counter.

reference lasers for every 100 GHz. The use of C_2H_2 or HCN absorption lines would be preferable for reference laser stabilization in the 1.5 µm region.

5.2 MULTICHANNEL FREQUENCY STABILIZATION

Hiromu Toba

Optical frequency stabilization is indispensable for the construction of optical FDM transmission systems. In particular, frequencies of the transmitters can be relatively stabilized to a common frequency reference, when optical sources of FDM carriers are gathered in the central station, such as point-to-point transmission and information distribution systems described by Nosu, Toba and Iwashita (1987).

This section describes the requirements of frequency stabilization systems and compares the various multichannel frequency stabilization methods used to realize such systems.

5.2.1 Requirements

The requirements for multichannel stabilization technology are listed as follows:

- frequency stability;
- number of channels supported;
- independence from modulation scheme;
- adoptability for absolute frequency reference;
- small size and high reliability.

In order to consider the requirement of frequency stability, we consider the requirements of an FDM point-to-point transmission system that employs cascaded MZ filter-type multi/demultiplexers (see section 4.1) and intensity modulation with direct detection schemes. Calculated frequency detuning characteristics between signal frequency and each multi/demultiplexer are shown in Fig. 5.7, according to Toba, Inoue and Nosu (1986). Note that the temperature dependence of the frequency drift of the laser diode (LD) and the MZ filter, made of silica glass, is $\sim 10\,\text{GHz/K}$ and $\sim 1\,\text{GHz/K}$, respectively. The multi/demultiplexers

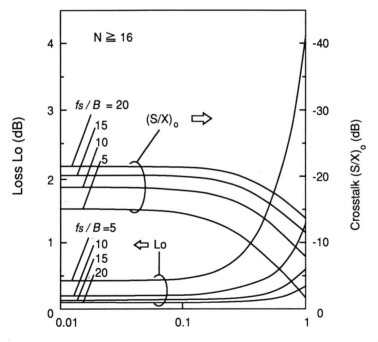

Fig. 5.7 Calculated frequency detuning characteristics for FDM system employing MZ filter multi/demultiplexer and IM/DD scheme.
Source: Toba, Inoue and Nosu, 1986.

are significantly more stable that the LDs. Note that temperature stabilization schemes can achieve fluctuations as small as 0.01K. The amount of frequency detuning, Δf, and the channel frequency spacing, f_s, are normalized by the transmission bit rate of the signals, B.

In Fig. 5.7, all laser frequencies are assumed to detune equally from the center frequency of the multiplexer and the demultiplexer. The transmission loss and the crosstalk of the multi/demultiplexers increase as the amount of detuning frequency increases. They are severely degraded as the frequency spacing becomes narrower. For example, the detuning frequency should be less than 0.5 and 0.2 times the bit rate for frequency spacings of 10 and 5 times the bit rate, respectively, to achieve a loss of 0.5 dB and a crosstalk of -12 dB. At the bit rate of 1 Gbit/s, they correspond to 500 MHz and 200 MHz for a frequency spacing of 10 and 5 GHz.

Although the required number of channels depends upon the application system, more than 100 channels can be transmitted by using the 1.55 µm fiber low-loss wavelength region. Modulation format independence is important for the stabilization system application. In order to improve the long-term stability and accuracy of the frequency reference, the ability to adopt an absolute frequency reference, which is described in section 5.1, is required.

5.2.2 Multichannel frequency stabilization method

Several methods for stabilizing multi-carriers have been proposed.

(a) Synchronous detection method

Each LD frequency is locked to an equally-spaced resonant frequency of a resonator, such as a Febry-Perot or ring resonator, by synchronous detection for direct FM or FSK signals, which are converted to AM or ASK signals by the resonator; see Glance, et al. (1987), Kaminow, et al. (1988), Toba, Oda and Nosu (1989) and Kuboki, et al. (1990).

(b) Beat pulse method/reference pulse method

Mixing the output of a tunable LD and multicarriers to make "beat pulses", then locking the generated time to the previously assigned reference time (beat pulse method) as in Bachus, et al. (1985). This method can be enhanced by using a Fabry-Perot resonator, which generates the equally-spaced reference pulse by passing a tunable LD light through the resonator (reference pulse method) described in Shimosaka, et al. (1990).

(c) Scanning interferometer method

Converting the frequencies of the FDM signals into time division multiplexed pulses by using a scanning Fabry-Perot interferometer and locking the time, on

MULTICHANNEL FREQUENCY STABILIZATION

which the pulses are generated, to the preassigned time as in Toba, *et al.* (1988) and Maeda, *et al.* (1989).

(d) Sideband locking method

LD frequencies are locked to sidebands that are generated by a phase-modulated He–Ne laser or a frequency-modulated LD as discussed in Hunkin, Hill and Stallard (1986) and Maeda and Kazovsky (1989).

Figure 5.8 shows the schematic of a typical synchronous detection method according to Toba, Oda and Nosu (1989). In this configuration, 16 FDM signals are stabilized to the resonant frequencies of a waveguide ring resonator with a free spectrum range of 5 GHz, and finesse of 12. The ring resonator is thermally stabilized to within 0.01 K, corresponding to a resonance frequency fluctuation of less than 15 MHz.

Frequency stabilization proceeds as follows. The frequencies of LDs are set in the vicinity of the resonant frequencies of the ring resonators. Each LD is directly frequency modulated at a different frequency $f_1, f_2, \ldots, f_{16}$ with a frequency

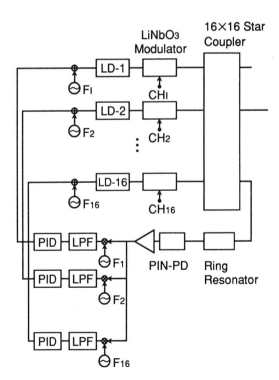

Fig. 5.8 Schematic of the frequency stabilization employing the synchronous detection method.
Source: Toba, *et al.*, 1989.

142   OPTICAL FREQUENCY DIVISION MULTIPLEXING SYSTEMS

deviation of Δf (pilot signals). The modulation frequency f_i ($i = 1,\ldots, 16$) is set at several kilohertz. The frequency fluctuation of each LD is converted to an amplitude fluctuation by means of the resonance peak. The output of the ring resonator is detected by a photo diode (PD). Each frequency fluctuation of the LD is extracted by the synchronous detection with a reference signal with a frequency of f_i through the mixer and the low pass filter (LPF). The fluctuation component is fed back to the bias current of each LD through the PID amplifier. As a result, each frequency is locked to a resonance peak of the ring resonator with 5 GHz spacing.

Time dependencies of the frequency fluctuations of the frequency-locked LDs ($f_5 - f_{10}$) are shown in Fig. 5.9. The frequency fluctuation in the case was approximately 200 MHz during free running, while it was reduced to less than 30 MHz for all 16 LDs when the stabilization was performed.

When frequencies of the pilot FM signals are separated at 200 kHz intervals, a 20 kHz bandwidth is required for multiplexing 100 optical carriers and FM frequencies can range from 20 kHz to 40 kHz. This frequency range is low enough not to degrade the transmission performance at a bit rate of more than several hundreds of Mbit/s. Therefore, 100 frequencies can be successfully stabilized simultaneously by means of this stabilization technique.

A schematic configuration of the reference pulse method is shown in Fig. 5.10 according to Shimosaka, et al. (1990). The frequency of the reference LD is swept periodically. The output is divided into two parts: one is coupled to the optical resonator, the other is combined with FDM signals and the beat pulses generated from the photo detector. Reference pulses are generated from the output of the optical resonator with a repetition corresponding to the FSR of the cavity. The control circuit monitors the generation time differences between reference pulses and beat pulses. The time differences correspond to the frequency differences between LD frequencies and the resonant frequencies of the resonator. As a

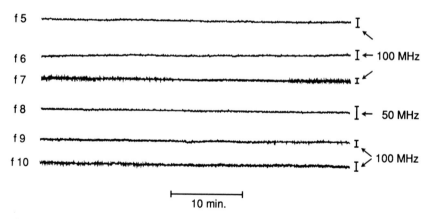

Fig. 5.9 Time dependence of the frequency fluctuation of stabilized laser diodes. Source: Toba, et al., 1989.

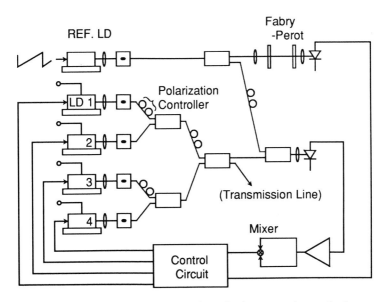

Fig. 5.10 Schematic configuration of reference pulse method.
Source: Shimosaka, et al., 1990.

result, each LD frequency coincides with the resonant frequency by means of the current feedback to the LD bias.

The frequency fluctuation of the four-channel LDs under frequency stabilization is shown in Fig. 5.11. Each LD fluctuation is suppressed to within 20 MHz to 150 MHz, while that of the free running state is 700 MHz.

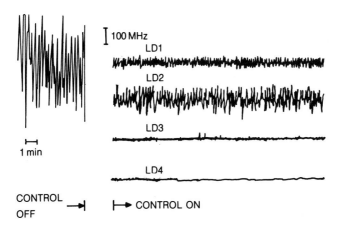

Fig. 5.11 Frequency fluctuation of the four-channel LDs under frequency stabilization employing the reference pulse method.
Source: Shimosaka, et al., 1990.

The channel capacity of the method depends on the continuous tunable range of the tunable LD. For example, 100 LDs are simultaneously stabilized at a frequency spacing of 5 GHz utilizing the tunable LD with a tunable range of more than 500 GHz. This method is available for any type of modulation scheme (Shimosaka, *et al.*, 1990).

Channel capacity for frequency stabilization depends on the frequency stabilization scheme. That of the scanning interferometer method depends on the finesse of the Febry-Perot interferometer. The number of sidebands determines the channel capacity of the sideband locking method. Long-term stability or absolute frequency accuracy is improved by adopting the absolute frequency stabilized references as in Maeda, *et al.* (1989).

5.3 FIBER NONLINEAR EFFECTS IN OPTICAL FDM SYSTEMS

Nori Shibata

Optical frequency division multiplexing (FDM) techniques have been intensely studied for future lightwave communication networks. A 100-channel optical FDM transmission/distribution experiment (Toba, *et al.*, 1990) has been demonstrated, and common amplification of 100 channels has also been achieved by employing an Er^{3+}-doped fiber amplifier in Inoue, Toba and Nosu (1991).

The advent of the optical amplifier raises the concern of optical nonlinear effects, and removes fiber-loss as the dominant concern in lightwave communication systems. The detrimental feature of fiber nonlinearities in multichannel transmission systems is intermodulation between signal channels. The problem of the SRS-induced crosstalk will be seen in FDM transmission systems with hundreds of channels (Chraplyvy, 1990). SBS-induced crosstalk will be of concern in bidirectional multichannel systems, even if transmitter powers injected into a fiber do not reach the critical level according to Waarts and Braun (1985). In the XPM process, the phase of the first channel is modulated by the intensity fluctuations of other channels propagating in a single-mode fiber as in Chraplyvy (1990). The FWM-induced crosstalk arises when system operation wavelengths are allocated very near to the zero chromatic dispersion wavelength according to Shibata, Braun and Waarts (1987).

Figure 5.12 shows an indication of transmitter power limitations imposed by the fiber nonlinearities in the 1.55 μm-wavelength region for a conventional non-dispersion-shifted (NDS) fiber with a zero chromatic dispersion wavelength of 1.3 μm. These results suggest that SRS and XPM are dominant concerns in multichannel systems employing NDS fibers and about 100 channels, and that the performance of multichannel systems with 10 channels is dominantly influenced by FWM and SBS.

Taking into consideration that the maximum transmitter power for the FWM process drastically varies with Δf and D_c which affect phase-matching conditions

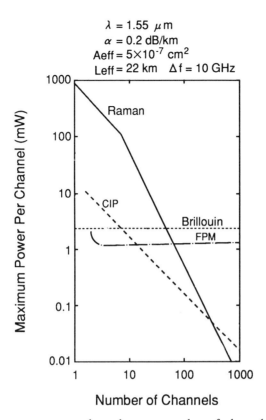

Fig. 5.12 Maximum power per channel versus number of channels. CIP and FPM stand for carrier-induced phase modulation and four-photon mixing, respectively. CIP is called self-phase modulation or cross-phase modulation. FPM means FWM in this text. Source: Chraplyvy, 1990.

in the FWM process, the maximum allowable power in the zero dispersion wavelength region is very important for a high-speed transmission system to investigate receiver sensitivity degradation due to FWM. FWM efficiency η can be reduced so as to become zero by setting the channel spacing greater than 100 GHz at 1.55 μm wavelength for a fiber length $L = 100$ km and $D_c = 1$ ps/km-nm as in Shibata, Braun and Waarts (1987). However, the efficiency η holds 100% over about 10 nm in the case of channel frequency allocation shown in Fig. 3.20 as recorded by Inoue and Toba (1992b). The worst case transmitter power limitation for the channel allocations is illustrated in Fig. 3.20.

A calculated result of maximum power per channel as a function of number of channels is shown in Fig. 5.13 for the channel allocations at a channel spacing of 10 GHz, fiber length of 50 km, and crosstalk level of -20 dB. The maximum power per channel for 101 channels is found to be approximately -14.5 dBm (~ 35 μW) which suggests that FWM becomes a dominant concern in multi-

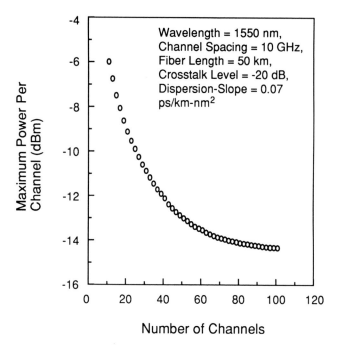

Fig. 5.13 Maximum power per channel as a function of number of channels for the channel arrangements shown in Fig. 5.17.

channel systems employing dispersion-shifted (DS) fibers and a number of channels up to a few hundreds. The results shown in Figs. 5.12 and 5.13 are for unmodulated optical carriers. Nevertheless, they indicate that FWM is expected to result in a serious limitation in multichannel systems with closely spaced channels.

5.3.1 FDM transmission system configuration

Schematic diagrams of optical FDM systems are illustrated in Fig. 5.14. Figure 5.14(a) shows the basic FDM configuration as an alternative to a high-speed time division multiplexing (TDM) system as in Shibata, et al. (1990). Figure 5.14(b) shows point-to-point FDM configurations applied to information distribution networks such as CATV networks. The descriptions of FDM in Fig. 5.14 represent optical frequency combiners, and OFSR denotes an optical frequency selective receiver utilizing a heterodyne detection (Bachus, et al., 1985a and Glance, et al., 1987) or a tunable optical filter (Toba, et al., 1988 and Kaminow, 1990).

The point-to-point configuration is applicable to optical FDM systems for long-haul transmission, and the point-to-multipoint configurations are applicable

FIBER NONLINEAR EFFECTS IN OPTICAL FDM SYSTEMS   147

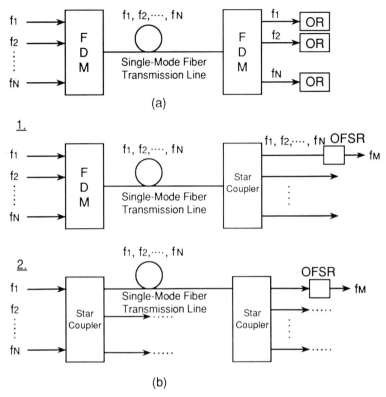

Fig. 5.14 Schematic diagrams of optical systems: (a) point-to-point transmission; (b) point-to-point transmission system.

to systems with transmission lengths less than several tens of kilometers. SBS, XPM, and FWM will be of dominant concern due to the restriction of transmitter power in the FDM for long-haul transmission shown in Fig. 5.14(a), and SRS and FWM will be considered in the FDM transmission systems shown in Fig. 5.14(b). As an example, the critical power levels of SBS in a CPFSK transmission system are described in section 3.4.1(b). Unlike the maximum power restricted by SBS, the limitation on transmitter power due to SRS or FWM should be determined by taking into account the amount of induced crosstalk between signals.

Multichannel system performance limits are estimated by calculating the depletion of the shortest wavelength channel in the Raman-induced crosstalk problem, while the amount of crosstalk depends on FWM efficiency related to chromatic dispersion, channel spacing, and fiber length as detailed in section 3.4.2(a). Specifically, chromatic dispersion characteristics are very important for evaluating the amount of FWM-induced crosstalk. Focusing on the dispersion

148   OPTICAL FREQUENCY DIVISION MULTIPLEXING SYSTEMS.

characteristics of single-mode fibers, NDS and DS fibers are now commercially available as low-loss single-mode transmission lines.

In the following sections, the induced crosstalks due to SRS, XPM, and FWM in multichannel systems are described. In general, the SBS effect does not couple channels in an FDM system.

5.3.2 SRS-induced crosstalk

As mentioned in section 3.3, the short wavelength channel acts as a pump wave for longer wavelength channels in WDM transmission systems. The SRS-induced crosstalk was analyzed by assuming the actual Raman gain profile of silica as approximated by a triangular function as shown in Fig. 5.15. Broken and solid curves represent the actual Raman gain profile and the triangular function, respectively.

Considering the case that all the channels fall within the Raman gain profile and equal signal power P (watts) is injected into an optical fiber, the fractional

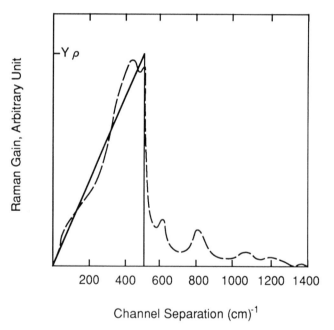

Fig. 5.15 Actual Raman gain profile (broken curve) and approximation used in calculations (solid curve).
Source: Chraplyvy, 1984.

FIBER NONLINEAR EFFECTS IN OPTICAL FDM SYSTEMS

power P_F lost by the shortest wavelength channel is written in Chraplyvy (1984) as:

$$P_F = \frac{g_R \Delta v L_{eff} P}{3 \times 10^{13} A_{eff}} \frac{N(N-1)}{2} \tag{5.6}$$

where N is the number of equally spaced channels and Δv is the channel spacing. The power penalty X_{SRS} due to SRS is written as:

$$X_{SRS} = -10 \log(1 - P_F) \, [\text{dB}] \tag{5.7}$$

The maximum allowable transmitter power per channel can be estimated by equations 5.6 and 5.7 for the required value of power penalty. For example, the maximum power per channel is 3 mW in a 10-channel system with $\Delta v = 1.3$ THz at 1.5 μm wavelength for the power penalty of 0.5 dB as in Chraplyvy (1984). The fractional power P_F is found from equation 5.6 to be proportional to channel spacing Δv. Therefore, Raman-induced crosstalk can be reduced by closely spacing the optical channels.

The influence of SRS will be seen in multichannel transmission systems with hundreds of channels, as is found from Fig. 5.12. Indeed, no power penalty due to SRS-induced crosstalk was observed in a 10-channel WDM system as in Olsson, et al. (1985).

5.3.3 XPM-induced crosstalk

When the phase-sensitive detection technique is employed, the error rate performance of the detection system will be influenced by XPM. In a PSK homodyne system employing phase modulators such as $LiNbO_3$ and semiconductor modulators, the residual amplitude modulation induced by the modulators degrades transmission performance as in Norimatsu and Iwashita (1991). The influence of XPM on a two-channel 10 Gbit/s PSK homodyne transmission system has been experimentally studied by using 100 km-long DS fibers at 1554 nm wavelength.

Figure 5.16 shows power penalty as a function of fiber input power at a BER of 10^{-8}. Open circles and a triangle represent the measured data for the BER experiments utilizing the respective phase modulators with and without anti-reflection coating at the adjacent channel. A power penalty was observed for more than 6 dBm fiber input power with 28% residual amplitude modulation, and power penalty due to SPM was not observed. The influence of SPM will be seen in in-line optical amplifier systems reported in Saito, et al. (1991). Possible origins of intensity changes are the intensity-modulated signal itself in IM-DD schemes and PM-to-AM or FM-to-AM conversion in coherent schemes.

The combined action of SPM and fiber dispersion, and the amplifier noise, have been intensely investigated in long-haul transmission systems at high bit rates for IM-DD schemes and coherent schemes in Saito, et al. (1991), Gordon

150 OPTICAL FREQUENCY DIVISION MULTIPLEXING SYSTEMS

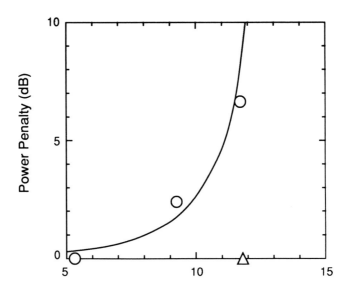

Fig. 5.16 Power penalty dependence on fiber input power. Power penalty was measured at a bit error rate of 10^{-8}. The values measured by using phase modulators with and without AR-coated facet at the adjacent channel are plotted with ($\triangle$) and ($\bigcirc$) respectively. The calculated values are shown by the solid curve.
Source: Norimatsu and Iwashita, 1991.

and Mollenauer (1991), Marcuse (1991), Hamaide and Emphlit (1990) and Hamaide, et al. (1992).

5.3.4 FWM-induced crosstalk

FWM-induced crosstalk is seen for the channel allocation shown in Fig. 3.19. When the total number of optical channels, which are equally spaced and have equal optical power of P_S, is $N = 2K + 1$ (K an integer) in an optical FDM system, total optical power P_{FWM} generated through the FWM process at a certain channel M of optical frequency f_M is given as the sum of all generated powers P_{ijk} of the frequency component f_{ijk} for all relevant combinations; that is, i, j and k take the values from 1 to $2K + 1$. The power P_{FWM} at the frequency f_M for unmodulated waves is expressed in Shibata, et al. (1990) and Waarts and Braun (1986) as:

$$P_{FWM}^{(CW)} = \frac{1024\pi^6 (\chi_{1111}^{(3)})^2 L_{eff}^2}{n^4 \lambda^2 c^2 A_{eff}^2} P_S^3 \exp(-\alpha L)(\Sigma D\eta_{ijk}) \qquad (5.8)$$

where η_{ijk} is the wave generation efficiency which is given by equation 3.43a. The amount of crosstalk between channels is defined as:

$$C = \frac{P_{FWM}^{(CW)}}{P_S} \tag{5.9}$$

The crosstalk level C can be calculated by using the values of chromatic dispersion D_c, channel separation Δf, fiber length L, and other parameters included in equations 3.42 to 3.43a and 5.8. Conversely, the maximum allowable power restricted by FWM can be evaluated from equations 5.8 and 5.9 by using the given C-value (i.e. $C = -20$ dB).

The amount of crosstalk C as a function of N, C as a function of fiber length L, and the maximum allowable power P_{max} as a function of N at a center channel are shown in Fig. 5.17 given in Shibata, et al. (1990). The values used for chromatic dispersion are $D_c = 1$ and 17, and $D_c = 23$ and 1 ps/km-nm at $\lambda = 1300$ and 1550 nm for NDS and DS fibers, respectively. The NDS fiber operated at $\lambda = 1550$ nm is found to keep multichannel transmission characteristics stable against FWM-induced crosstalk.

Receiver sensitivity degradation due to FWM was the first to be experimentally verified in a three-channel heterodyne transmission system reported in Shibata, et al. (1990) and Shibata, Iwashita and Azuma (1989). The experimental setup for measuring the receiver sensitivity degradation is illustrated in Fig. 5.18. LD_1, LD_2 and LD_3 are DFB lasers operated at 1550 nm. The LD_1 and LD_2 of the respective frequencies f_1 and f_2 were used in CW operation to generate nonlinear waves due to FWM, and the LD_3 of frequency f_3 was used as a transmitter laser modulated by a 2 Gbit/s NRZ signal. The 2 Gbit/s CPFSK signal with 1.4 GHz frequency deviation was obtained by the modulation scheme.

Light beams from the three laser diodes were launched into a test fiber which was an 18 km-long NDS fiber or a 26 km-long DS fiber having a dual-shape core profile as shown in Fig. 3.10. The transmitted signal was heterodyned with the wave of a local DFB laser. The channel frequency allocation is also shown in the inset of Fig. 5.18. The three frequencies were equally spaced at a frequency separation Δf; then the generated wave of $f_{221} = 2f_2 - f_1$ overlapped the transmitter laser frequency in the same frequency region. The IF bandwidth ranged from 1.5 to 5.5 GHz at an intermediate frequency of 3.5 GHz. The FWM efficiency η_{221} for the NDS fiber approached close to zero at $\Delta f = 20$ GHz, while η_{221} for the DS fiber held approximately 100%. The bit error rate (BER) measurements were carried out at $\Delta f = 21.6$ GHz because of the $\eta_{221} \sim 0\%$ and 100% for NDS and DS fibers, respectively.

Figure 5.19 shows BER as a function of average received signal power for the input CW powers of $P_1 = 0.3$ and $P_2 = 1.3$ mW. The measured receiver sensitivity degradations were 0.5 and 4 dB for the respective NDS and DS fibers. In the inset of Fig. 5.19, eye patterns observed at a signal power $P_3 = -38$ dBm, with and without FWM, are also shown. It is clear that the NDS fiber is more suitable

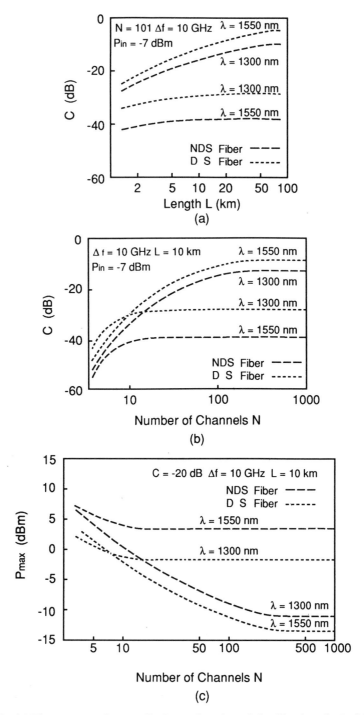

Fig. 5.17 (a) The amount of crosstalk C as a function of the fiber length, L; (b) C as a function of channels, N; (c) the maximum allowable transmitter power injected into a fiber as a function of N.
Source: Shibata, *et al.*, 1990.

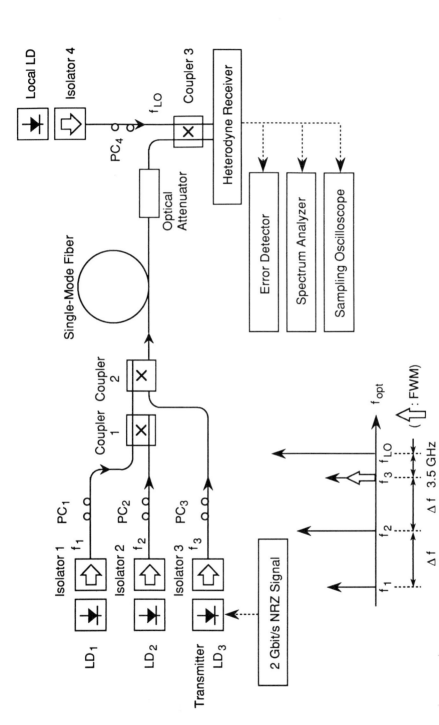

Fig. 5.18 Experimental setup for measuring the receiver sensitivity degradation due to the FWM process in a three-channel system. Source: Shibata, et al., 1990.

154 OPTICAL FREQUENCY DIVISION MULTIPLEXING SYSTEMS

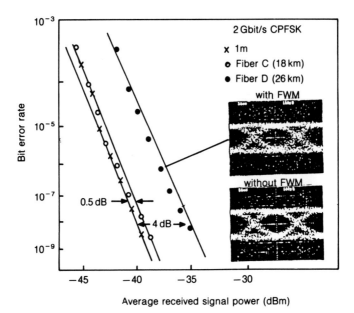

Fig. 5.19 Bit error rate measurement for fibers C and D and the observed eye patterns with and without the lightwave generated through the FWM process.
Source: Shibata, et al., 1990.

for FDM systems operating at $\lambda = 1550$ nm, since the generated power due to FWM rapidly decreases with Δf according to chromatic dispersion values around $15 \sim 20$ ps/km-nm at this wavelength.

Similar BER experiments were demonstrated in a 16-channel FSK-AMI heterodyne detection system by Maeda, et al. (1990), and in a 4-channel FSK direct detection system by Inoue and Toba (1991) and Inoue, Toba and Oda (1992). The crosstalk from multiple FWM waves was studied for the two cases of laser diodes with CW operation and with FSK modulation in the 16-channel heterodyne system in Maeda, et al. (1990).

Figure 5.20 shows the BER as a function of the detected power of the laser of channel 7 which is located near the center channels 8 and 9 in a 16 channel system. A sensitivity degradation of 3 dB was measured when only laser 7 was modulated and the other lasers were operated in the CW mode. When all 16 lasers were modulated, the sensitivity penalty was smaller at 1.8 dB. This is because the FWM spectrum spreads when the interacting signals are modulated; hence the crosstalk is reduced as a result and the receiver IF filter captures less interfering noise.

The influence of fiber FWM on multichannel FSK heterodyne envelope detection systems has been theoretically investigated by Inoue and Toba (1992a). The power penalty X_{FWM} is given by:

$$X_{FWM} = -10\log\left[1 - \left(\frac{S}{N}\right)_0 \left(\frac{P_{FWM}^{(FSKH)}}{P_S}\right)\right] \tag{5.10}$$

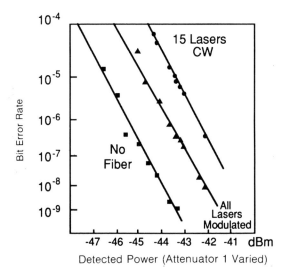

Fig. 5.20 Bit error rate (BER) measurement on laser 7. Attenuator 1 was varied to make this measurement. (Squares: no fiber; triangles: all lasers modulated; circles: 15 lasers CW.) Source: Maeda, *et al.*, 1990.

where $(S/N)_0$ is the signal-to-noise ratio at the IF stage of the "mark" frequency band without FWM, $P_{\text{FWM}}^{(\text{FSKH})}$ is the sum of generated power for non-degenerate and partially-degenerate channel combinations, and P_S is the received power to obtain a given BER (i.e. BER = 10^{-9}). Comparing the sums of the generated power due to FWM for the two cases of unmodulated waves and FSK modulated waves,

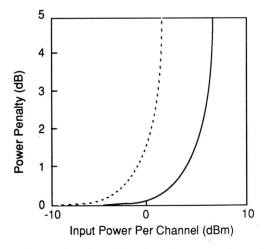

Fig. 5.21 Power penalty as a function of fiber input power per channel. Solid and broken curves are calculated results for nondispersion-shifted and dispersion-shifted fibers, respectively. Fiber length is set at 50 km.

$P_{\text{FWM}}^{(\text{FSKH})} \approx \frac{3}{8} P_{\text{FWM}}^{(\text{CW})}$ is valid because the non-degenerate channel combination mainly contributes to the sum of the generated power according to Inoue and Toba (1992). As an example of the calculated result for a 4-channel system with a channel spacing of 10 GHz, the power penalty as a function of fiber input power per channel is shown in Fig. 5.21. Parameter values used in the calculation are chromatic dispersion $D_c = 17$ and 1 ps/km-nm, fiber attenuation coefficient $\alpha = 0.2$ and 0.27 dB/km, and mode field diameter $2W = 11$ and 8.5 µm for 50 km-long NDS and DS fibers, respectively, as $(S/N)_0 = 80$ for heterodyne envelope detection (Okoshi, et al., 1981). The power penalty drastically increases with an input power greater than 0 and 5 dBm, and the maximum allowable powers at $X_{\text{FWM}} = 0.5$ dB are found to be approximately -2 and 3 dBm for NDS and DS fibers, respectively.

In the comparison of receiver sensitivity degradations due to FWM between the 4-channel FSK heterodyne envelope detection and FSK direct detection systems, it has been reported that the FSK heterodyne detection suffers more from FWM influence than an FSK direct detection scheme according to Inoue, Toba and Oda (1992) and Inoue and Toba (1992).

5.4 CHANNEL SELECTIVE RECEIVER UTILIZING HETERODYNE DETECTION

Hiromu Toba

One of the advantages of the coherent detection technique is that it can select and receive one channel from among optical frequency division multiplexed signals utilizing the tunable local oscillator (LO) and the intermediate frequency (IF) filter (DeLange, 1970). Research into coherent multichannel systems utilizing tunable laser diodes began in 1984 according to Bachus, et al. (1984). Much information, both theoretical and experimental, has been reported since and is considered to be applicable to the information distribution systems as well as to the increase of transmission capacity in the point-to-point transmission systems See references as under:

Kazovsky (1987a)
Agrawal (1987)
Kazovsky (1987b)
Suyama, Chikama and Kuwahara (1988)
Kazovsky and Gimlett (1988)
Kazovsky and Jacobsen (1989)
Elrefaie, Maeda and Guru (1989)
Cheng and Okoshi (1990)
Bachus, et al. (1986)
Bachus, et al. (1989)

Glance, *et al.* (1988a)
Glance, *et al.* (1988b)
Glance and Scaramucci (1990)
Park, *et al.* (1988)
Shibutani, *et al.* (1989)
Welter, *et al.* (1989)
Yamazaki, *et al.* (1990a)
Flaarønning, *et al.* (1990)
Noé, *et al.* (1990)
Yamazaki, *et al.* (1990b)
Stanley, Hill and Smith (1987)
Fujiwara, *et al.* (1990)
Yamazaki, *et al.* (1990c)
Tsushima, *et al.* (1991).

One of the key issues of the coherent multichannel transmission is to determine the optimum channel separation to the bit rate ratio for various modulation/demodulation schemes. Narrower channel spacing enables us to transmit the larger number of channels under a certain wavelength bandwidth which is restricted by the tunable range of the LO and/or gain spectral bandwidth of a common optical amplifier. The channel spacing of the heterodyne detection receiver is reduced by employing the image band rejection technique covered by Glance (1986) and Chikama, *et al.* (1990), as well as the phase diversity technique and homodyne detection, which are described in Chapter 3.

This section describes the principle of the coherent receiver for multichannel transmission, optimum channel separation, an image rejection receiver and a tunable laser and controller.

5.4.1 Principle

Figure 5.22 shows the configuration of a channel selective receiver utilizing heterodyne detection. Optical FDM signals are combined with LO and converted by the photodetector into an electrical signal. A specified channel signal is selected by the following intermediate frequency filter (IF filter) and demodulated by the demodulation circuit.

The frequency allocation of the FDM signals and LO light is shown in Fig. 5.23. Channel selection by the IF filter is done by positioning the LO frequency near the desired channel. In Fig. 5.23(a), frequency spacings of the FDM signals are set equal (f_s), and the LO frequency is positioned at a lower frequency to the desired channel (f_i) in the optical frequency domain. After heterodyne detection, the image band signal, which is on the low frequency side (f_{i-1}) of the LO frequency (f_{LO}) in the optical domain and converted to $F_{i-1} = f_s - (f_i - f_{LO})$ in the IF domain, is interleaved in between the desired channel frequency ($F_i = f_i - f_{LO}$)

158   OPTICAL FREQUENCY DIVISION MULTIPLEXING SYSTEMS

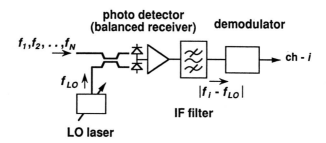

Fig. 5.22 Configuration of channel selective receiver utilizing heterodyne detection.

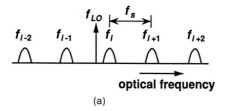

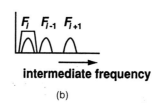

Fig. 5.23 Frequency allocation: (a) optical frequency domain; (b) intermediate frequency domain.

and next higher channel frequency ($F_{i+1} = f_{i+1} - f_{LO}$) in the IF domain as shown in Fig. 5.23(b). As a result, a large enough frequency interval is needed between channels to avoid interference from the image band signal (F_{i-1}) by the IF filter.

We now consider the balanced coherent receiver for use in a multichannel system. The configuration of the balanced receiver is shown in Fig. 5.22. The output currents of the photodetectors i_1, i_2 are according to Kazovsky (1987a):

$$i_1 = \frac{1}{2}\frac{\eta e}{h\nu}P_{LO} + \frac{1}{2}\frac{\eta e}{h\nu}\sum_{k=1}^{N}\sum_{l=1}^{N}E_k E_l^* + \frac{\eta e}{h\nu}\sqrt{P_{LO}P_S}m_K \cdot \cos(2\pi f_{IF}t + \Phi_N)$$

$$+ \frac{\eta e}{h\nu}\sqrt{P_{LO}P_S}\sum_{k=1;k\neq K}^{N}m_k \cdot \cos[2\pi f_{IF} + 2\pi(k-N)f_s t + \Phi_k] + n_1(t) \quad (5.10)$$

and

$$i_2 = \frac{1}{2}\frac{\eta e}{h\nu}P_{LO} + \frac{1}{2}\frac{\eta e}{h\nu}\sum_{k=1}^{N}\sum_{l=1}^{N}E_k E_l^* - \frac{\eta e}{h\nu}\sqrt{P_{LO}P_s}m_K \cdot \cos(2\pi f_{IF}t + \Phi_N)$$

$$- \frac{\eta e}{h\nu}\sqrt{P_{LO}P_s}\sum_{k=1;k\neq K}^{N}m_k \cdot \cos[2\pi f_{IF} + 2\pi(k-N)f_s t + \Phi_k] + n_2(t) \quad (5.11)$$

where P_{LO} is the power of the LO, P_s is the peak signal power, m_k and Φ_k reflect the possible amplitude and angle modulation of the kth channel, f_{IF} is the IF frequency, K is the subscript of the desired channel, N is the total number of multiplexed channels, and $n(t)$ is the equivalent noise current. In equations 5.10 and 5.11, the first term is the DC current due to the LO, the second term represents the direct detection and channel/cross-channel interference currents, the third term is the desired signal, the fourth term is the LO/undesired channel interference current, and the last term is the noise. The total current of the differential output of the photodetectors according to Kazovsky is:

$$i_T = i_1 - i_2$$
$$= A\left\{m_K\cos(2\pi f_{IF}t + \Phi_k) + \sum_{k=1;k\neq K}^{N}m_k\cos[2\pi f_{IF}t + 2\pi(k-K)f_s t + \Phi_k]\right\} + n(t)$$
$$(5.12)$$

where

$$A = 2\frac{\eta e}{h\nu}\sqrt{P_{LO}P_s}$$

and $n(t) = n_1(t) - n_2(t)$. The first term is the desired signal and the second term is the LO/undesired channel interference current. Equation (5.12) shows that the direct detection terms and the channel/cross-channel interference terms are not present at the output of the balanced receiver and, therefore, do not deteriorate the performance of the receiver. The power spectral density (PSD) of i_T, $G_T(f)$, is:

$$G_T(f) = A^2[G_s(f) + G_c(f)] + \xi(f) \quad (5.13)$$

where $G_s(f)$, $G_c(f)$ and $\xi(f)$ are the PSD of the signal of the desired channel, the crosstalk from the undesired channel and $n(t)$, respectively.

If we approximate the crosstalk probability density function by the Gaussian probability function, the bit error rate (BER) according to Kazovsky and Gimlett (1988) is:

$$\text{BER} = Q(\gamma_{TOT}) \quad (5.14)$$

where Q is the Marcum Q-function and

$$\gamma_{TOT}^2 = \frac{1}{\dfrac{1}{\gamma^2}+\dfrac{1}{\gamma_{CT}^2}} \tag{5.15}$$

γ_{TOT} is the total signal-to-noise ratio including shot noise, thermal noise, and crosstalk; γ is the signal-to-noise ratio including the shot noise and the thermal noise, and γ_{CT} is the signal-to-noise ratio for the crosstalk. γ and γ_{CT} are defined as follows.

$$\gamma \equiv \frac{\mu_1 - \mu_0}{2\sqrt{\sigma_t^2 + \sigma_s^2}} \tag{5.16}$$

$$\gamma_{TOT} \equiv \frac{\mu_1 - \mu_0}{2\sigma_c} \tag{5.17}$$

where μ_1 is the mean value of the signal at the decision circuit when binary "1" is transmitted, μ_0 is the mean value of the signal at the decision circuit when binary "0" is transmitted, σ_s^2, σ_t^2 and σ_c^2 are the variances of the shot noise, the thermal noise and the crosstalk, respectively. σ_c^2 is given by:

$$\sigma_c^2 = A^2 \int_{-\infty}^{\infty} G_s(f)|H(f)|^2 df \tag{5.18}$$

where $H(f)$ is the transfer function of the demodulator.

The receiver sensitivity penalty at BER $= 10^{-9}$, where the argument of the Q-function has to be 6, is:

$$\text{Penalty, dB} = -10\log\left(1 - \frac{36}{\sigma_{CT}^2}\right) \tag{5.19}$$

The calculated sensitivity penalty for various modulation formats as a function of normalized electrical domain channel spacing D_{nel} is shown in Fig. 5.24. Note that a matched filter is used as the demodulator.

If the smallest IF has been selected, the optical domain channel spacing of f_s is obtained as:

$$f_s = D_{nel} \cdot B + 2f_{IF} \tag{5.20}$$

where B is the bit rate. Imperfect balance of the balanced receiver, intentional or unintentional filtering of the data and/or of the transmitted signal, and the influence of laser phase noise are not taken into consideration in this calculation.

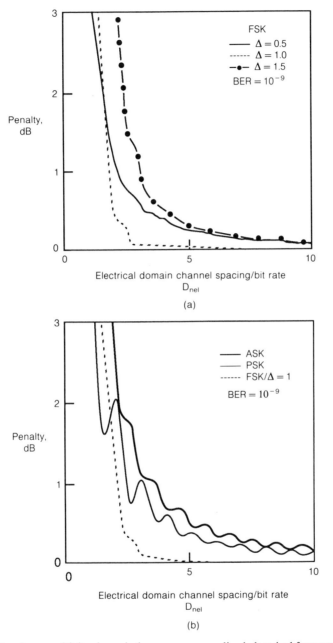

Fig. 5.24 Receiver sensitivity degradation versus normalized electrical frequency domain spacing D_{net} for various modulation/demodulation schemes. Note that the optical frequency spacing f_s is given as $f_s = D_{net}B + 2f_{IF}$. (a) FSK systems with various modulation indexes; (b) ASK, PSK and FSK systems; for FSK, the case $\Delta = 1$ is shown.
Source: Kazovsky and Gimlett, 1988.

Although the calculation is carried out in time domain analysis, the frequency domain analysis, which is described above, also gives similar results according to Kazovsky and Gimlett (1988). For FSK systems, the required channel spacing for a modulation index $\Delta = 1$ is narrower than that of $\Delta = 0.5$ and $\Delta = 1.5$. Normalized electrical channel spacing is required to be 3.3, 2.2 and 4.1 for $\Delta = 0.5$, 1.0 and 1.5 to keep a sensitivity penalty less than 0.5 dB. Both ASK and PSK systems require larger channel spacings than FSK systems with $\Delta = 0.5$ and 1.0.

ASK and PSK systems require similar normalized electrical channel spacings for a sensitivity penalty less than 0.5 dB; they are 4.8 for ASK systems and 4.2 for PSK systems. The IF frequency is usually set to be $f_{IF} = 2B$ according to Iwashita and Takatio (1990); then the optical frequency spacing is required to be more than $6.2B$ for FSK systems with $\Delta = 1.0$ and more than $8.8B$ for ASK and $8.2B$ for PSK systems.

5.4.2 Channel spacing reduction by the image rejection receiver

As mentioned in section 3.5.4 the image rejection receiver (IRR) can suppress the image signal in the IF domain. When the IRR is adopted in the coherent multichannel system, the crosstalk from the image signal is suppressed, and we can allocate the FDM channel signals with narrower spacing than the conventional heterodyne detection described previously (see Fig. 5.23).

Figure 5.25 shows the frequency allocation when the IRR is adopted in the FDM system. If the image band signal of the adjacent channel is suppressed enough, we can set the optical frequency spacing, f_s, equal to the electrical

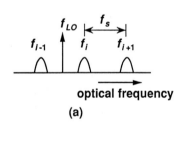

optical frequency

(a)

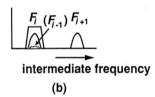

intermediate frequency

(b)

Fig. 5.25 Frequency allocation of FDM systems deploying the image rejection receiver: (a) optical frequency domain; (b) intermediate frequency domain.

frequency spacing and it is given as

$$f_s = D_{nel} \cdot B \tag{5.21}$$

According to 5.4.1, the optical channel frequency for FSK with $\Delta = 1.0$, ASK and PSK formats are set as $2.2B$, $4.8B$, $4.2B$ to keep the power penalty less than 0.5 dB.

Naito, et al. (1989) performed a two-channel DPSK system experiment with a bit rate of 560 Mbit/s using the IRR. Figure 5.26 shows a schematic of the experimental setup: 1.54 μm external fiber cavity DFB laser (EFC-DFB-LD) modules are used as transmitters and local oscillators (LO). The two signal lightwaves were independently modulated by the $LiNbO_3$ phase modulators. After the signals are transmitted through the fiber, they are detected by the IRR with manually compensated polarization. One output from the IRR is the real-band signal ($f_{s1} > f_L$), while the other output is the image band signal ($f_{s2} > f_L$). The real-band signal is demodulated by a differential detection scheme with a one-pulse delay circuit. The IF frequency is set to be four times the bit rate.

Figure 5.27 shows crosstalk penalties as a function of the frequency allocation of the interference channel f_{s2} normalized by the bit rate. Even when the image-band signal overlapped the real-band signal in the IF domain, which is the case when the $f_{s2} - f_L$/bit rate is -4, the signal f_{s1} can be received using the IRR with a power penalty of less than 1.4 dB. This experiment confirms that the

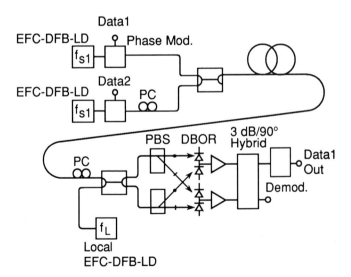

Fig. 5.26 Experimental setup of two-channel 560 Mbit/s DPSK heterodyne transmission experiment using the image rejection receiver.
Source: Chikama, et al., 1990.

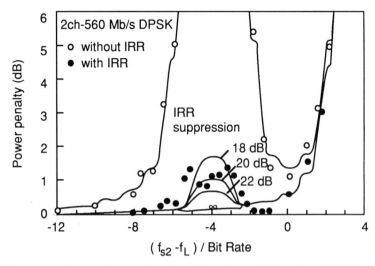

Fig. 5.27 Crosstalk penalties as a function of the channel allocation. Solid lines indicate the calculation results.
Source: Chikama, et al., 1990.

channel spacing can be set to three times the bit rate using the IRR with a 1.4 dB penalty, while it must be set to 11 times wider than the bit rate without the IRR.

Although the conventional IRR is sensitive to the polarization fluctuation of the transmitted signals, it will be improved by adopting the polarization diversity scheme as in Chikama, et al. (1990).

5.4.3 Tunable lasers for heterodyne detection

When the coherent detection receiver selects an arbitrary channel from among frequency division multiplexed channels, the laser local oscillator with its wide tuning range is indispensable. Figure 5.24 reveals that a channel spacing of about 10 times the bit rate is required to avoid the sensitivity degradation due to the crosstalk from the adjacent channels. For example, the channel spacing of 10 GHz corresponds to the bit rate of 1 Gbit/s. With 10 GHz spacing, the tunable range of the LO for 100 channels is 1000 GHz, which corresponds to 10 nm at a wavelength of 1550 nm.

Table 5.3 summarizes the characteristics of the monolithic tunable lasers. Various types of tunable lasers are investigated for LO, such as multisection distributed Bragg reflector (DBR) lasers (Murata, Mito and Kobayashi, 1987, Kotaki, et al., 1987, Kobayashi and Mito, 1988), and multisection distributed feedback (DFB) lasers (Yoshikuni, et al., 1986, Kotaki, et al., 1989, and Kuznetsov, 1988). Continuous tuning characteristics of the three-section DBR laser diode are

Table 5.3 Characteristics of monolithic tunable lasers

Laser	Tuning range		Linewidth (MHz)	References
	(quasi-)continuous (nm)	Discrete (nm)		
Multisection DBR	4.4	22	< 30	1, 2, 3
Multisection DFB	1.9	—	< 1	4
TTG	1.5 (7 nm predicted)	—	—	5, 6
Multisection interferometric	6	23	—	7

References
1. Murata, Mito and Kobayashi (1988)
2. Öberg, et al. (1991)
3. Yamazaki, et al. (1990)
4. Kotaki, et al. (1989)
5. Amann, et al. (1989a)
6. Amann, et al. (1989b)
7. Wuenstel et al. (1990)

shown in Fig. 5.28(a) as in Murata, Mito and Kobayashi (1988). This laser is segmented in the active, phase control and DBR sections with lengths of 290, 100 and 770 µm, respectively. In this case, the currents feeding to the phase control section and DBR section are changed simultaneously to maintain one DBR mode. Over 4.4 nm (550 GHz) a continuous tuning range at 1 mW light output is achieved with a sidemode suppression ratio of more than 30 dB.

The largest tuning ranges (discrete wavelength change) achieved so far with monolithic devices are ~ 20 nm, using a thermally-induced refractive index change (Öberg, et al., 1991) or multisection interferometric lasers (Wuenstel, et al., 1990). Recently, a significantly larger tuning range has been predicted for codirectional coupling between two different spatial modes of a twin-waveguide structure, which is called a tunable twin-guide (TTG) laser as in Amann, et al. (1989a) and Amann, et al. (1989b). The advantage of the laser is that the wavelength is controlled by only a single current, while the three-section DBR laser is a rather complicated mutual adjustment of the currents flowing into the phase control and the Bragg reflector sections, which is described above. The tuning range of the TTG laser is predicted up to 6 ~ 7 nm and preliminary experimental results report that the maximum continuous tuning range is 1.5 nm at a tuning current of 70 mA (Amann, et al., 1989b).

5.4.4 Channel selection technique

When the coherent optical FDM system is applied to the information distribution system, the heterodyne receiver should randomly receive an arbitrary channel

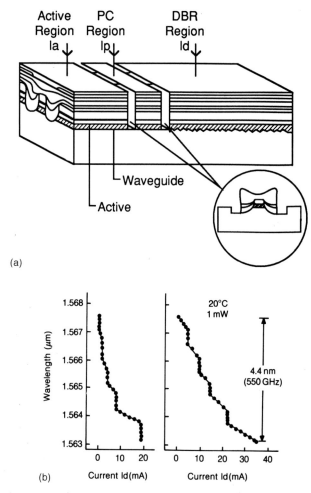

Fig. 5.28 Configuration and wavelength tuning characteristics of DBR laser diode: (a) configuration; (b) frequency tuning characteristics.
Source: Yamazaki, *et al.*, 1990.

from among optical FDM channels with a wide capture range and lock range. Figure 5.29 shows a schematic of the 10-channel coherent information distribution experimental setup according to Yamazaki, *et al.* (1990). The block on the left is an 8 GHz-spaced, 10-channel optical FDM transmitter with a channel spacing controller, which is described in section 5.2 (Shimosaka, *et al.*, 1990). Each laser is FSK modulated with a bit rate of 400 Mbit/s. Multisection DBR lasers are used in both transmitters and the LO in the receiver according to Murata, Mito and Kobayashi (1988).

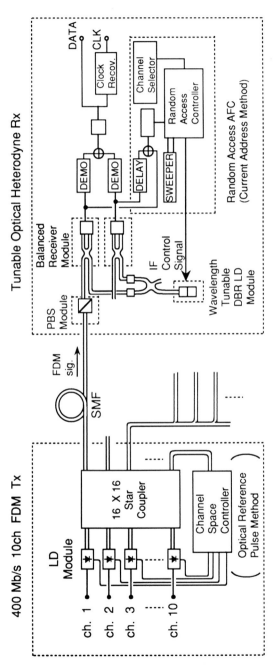

Fig. 5.29 10-channel optical FDM information distribution system employing random access optical heterodyne receiver. Source: Yamazaki, et al., 1990.

168 OPTICAL FREQUENCY DIVISION MULTIPLEXING SYSTEMS

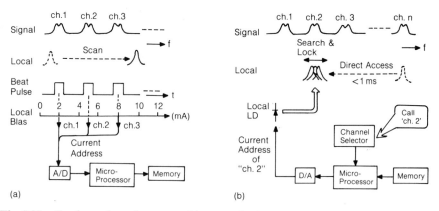

Fig. 5.30 Configuration of current address method. (a) Initialize and (b) channel select. Source: Yamazaki, *et al.*, 1990.

The block on the right is the random access optical heterodyne receiver with a polarization diversity scheme. A wideband capture range $(-f_{IF} \sim +f_{IF})$ of the AFC circuit is realized by employing the frequency discriminator, which utilizes the differential output of the low pass filter and high pass filter with the cut-off frequencies of f_{IF}. The random access controller operates according to the "current address method", which is shown in Fig. 5.30. In the initialization procedure, the controller scans the local frequency over the entire tuning range (over 100 GHz) and memorizes the local laser's current value, when the beat pulse is detected at the LPF in the IF discriminator. The memorized value corresponds to the current address for each channel. When a channel call is sent to the channel selector, the controller can directly access the memorized address current, which corresponds to the called channel. Subsequently, the sweeper scans the local frequency over the FDM signal bandwidth, searching for and locking the called channel IF into the capture range of the IF discriminator. A 114 GHz, wide capture range and 80 GHz lock range are obtained for the receiver.

These results are sufficiently wide for 10-channel FDM signals covering an 80 GHz total bandwidth. Random access selection is performed within 1 ms, which is determined by the control circuit response.

5.5 CHANNEL SELECTIVE RECEIVER UTILIZING OPTICAL FILTER AND DIRECT DETECTION

Hiromu Toba

A direct frequency-shift-keyed (FSK) laser transmitter has the advantages of avoiding the large modulating current and excessive frequency "chirp" which are inherent in direct intensity modulation. It is a simple transmitter configuration compared to the transmitter employing an external modulator. An optical FSK signal is converted to an ASK signal in the optical domain utilizing the optical filter as a discriminator, and the converted ASK signal is directly detected as in Saito, Yamamoto and Kimura (1982) and Malyon and Stallard (1989). This modulation/demodulation scheme is simple and cost-effective compared to the FSK/heterodyne detection scheme and is applicable to the optical FDM transmission because the multichannel signals can be densely spaced in the optical frequency region.

A Fabry-Perot filter and a Mach-Zehnder filter are available for the optical frequency discriminator (OFD). The Fabry-Perot filter has the functions of both the frequency discriminator and the optical demultiplexer, and the simple channel selective receiver configuration is feasible according to Kaminow, et al. (1987), Kaminow, et al. (1988), Willner, et al. (1990) and Willner (1990). The Mach-Zehnder filter with balanced receiver configuration has the advantage of the optical power efficiency in use (Toba, et al., 1990, and Toba, Oda and Nosu, 1991). This optical demodulation technique is used to operate a high speed single channel at 8 Gbit/s with a bulk FP (Chraplyvy, et al., 1989) and 11 Gbit/s (Vodhanel, et al., 1990) and 20 Gbit/s with bulk MZ (Shirasaki, Yokota and Touge, 1990).

This section describes the principle of the optical FSK-direct detection receiver (sub-section 5.5.1), design considerations (5.5.2) and transmission characteristics (5.5.3).

5.5.1 Principle

(a) Fabry-Perot filter/discriminator

Figure 5.31 shows the transmittance characteristics of the Fabry-Perot filter (FP filter) and frequency allocation of the optical frequency division multiplexed FSK signals according to Kaminow, et al. (1987 and 1988).

The power spectrum of the continuous-phase FSK signal consists of two peaks separated by the frequency deviation f_d with a large deviation index $\mu = f_d/B$. For $\mu \gg 1$, the FSK spectrum is approximately the superposition of two ASK spectra centered on the "1" and "0" frequency tones. The bulk of the energy in each peak is contained within a width equal to the bit rate B. The free spectral range (FSR) of the FP filter is set to be more than the bandwidth of the multiplexed signals, and the bandwidth of the FP filter is set to equal the bit rate B. A channel can be selected and the FSK signal converted to undistorted ASK format simultaneously by a bandpass filter of width B, tuned to the "1" (or "0") peak, placed in front of the PD. The modulation index μ must be sufficiently large to reduce crosstalk between "0"s and "1"s due to the overlap of the wings of their spectra. In the multichannel system, the wing of the "0" peak in the next-higher frequency channel will also introduce the crosstalk unless the channel spacing f_s is not sufficiently large.

170 OPTICAL FREQUENCY DIVISION MULTIPLEXING SYSTEMS

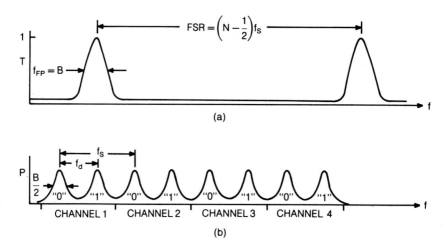

Fig. 5.31 (a) Transmittance characteristics of the Fabry-Perot (FP) filter. (b) Frequency allocation of the optical FDM FSK signals.
Source: Kaminow, *et al.*, 1988.

The analysis of the FP filter transmittance shows that the maximum channel density occurs for $f_s = 2f_d$ and that an FP filter characteristic with a $-3\,\text{dB}$ passband width of B allows a minimum spacing, $f_s = 6B$, with negligible (0.1 dB) power penalty according to Kaminow (1988). At tandem, the configuration of the FP filter allows a closer spacing, $f_s = 3B$.

The number of channels N is determined by:

$$N = \frac{\text{FSR}}{f_s} = \frac{F}{6} \tag{5.22}$$

in the ideal case when $f_s = 6B$, and $f_{FP} = B$. Note that N is independent of B and is determined only by the lineness F (Kaminow, 1990).

A tunable fiber Fabry-Perot (FFP) filter is promising for the FP filter. The configuration of the FFP filter is shown in Fig. 4.4(c). $F = 200$ with $5\,\text{dB}$ insertion loss is realized, and equation 5.22 shows that a 33-channel FDM system is possible. The FFP design is best suited to a length L in the 0.1–10 mm range, corresponding to an FSR in the range 10–1000 GHz.

(b) MZ filter/discriminator

The optical FDM FSK-direct detection scheme employing the Mach-Zehnder optical frequency discriminator (MZ-OFD) is shown in Fig. 5.32 as in Toba, *et al.* (1990) and Toba, Oda and Nosu (1991). In this case, one of the multiplexed FSK signals is selected by the optical filter before the optical frequency discrimination. Multistage MZ filters (periodic filters) are applicable for channel selection of the

OPTICAL FILTER AND DIRECT DETECTION 171

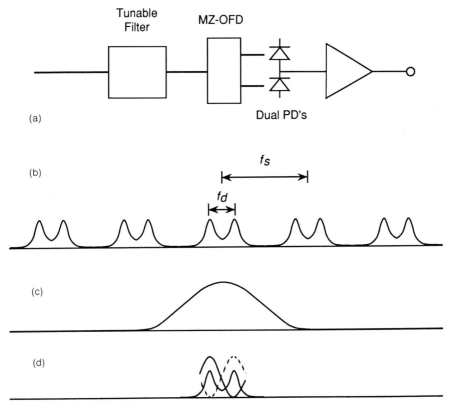

Fig. 5.32 (a) Configuration of FDM FSK direct detection scheme employing the Mach-Zehnder optical discriminator. (b) frequency allocation of the multiplexed FSK signals. (c) Transmittance of the frequency selection switch. (d) Frequency allocation of the selected FSK signal and transmittance of the MZ-OFD.

densely spaced multiplexed signals. The selected FSK signal is converted to an ASK signal by the MZ-OFD. The spacing between the minimum and maximum transmittance frequency of the MZ-OFD is set equal to the frequency deviation of the FSK signal. The center frequency of the optical signal is set at the crosspoint of the transmittance curves for the two output ports of the MZ-OFD. Then the mark component appears at one port and the space component appears at the other port. These outputs are differentially detected by photodiodes (PDs).

This configuration uses the transmitted power more efficiently than the single output configuration, such as FP-OFDs. Therefore, the receiver with a dual PD configuration is more sensitive than one with a single PD configuration by up to 3 dB. The following section focuses upon the receiver employing MZ-OFD with dual PDs.

5.5.2 Design considerations

The bit error rate (BER) for FSK-FDM direct detection systems with non-zero linewidth transmitters and dual APDs is described by Toba, Oda and Nosu (1991) as:

$$\text{BER} = \frac{1}{\sqrt{2\pi}} \int_Q^\infty \exp\left(-\frac{x^2}{2}\right) dx \quad (5.23)$$

with

$$Q = \frac{2\sqrt{S}}{\sqrt{\sigma_s^2 + \sigma_t^2 + \sigma_{\Delta v}^2 + \sigma_c^2}} \quad (5.24)$$

where S is received signal power, σ_s^2 is shot noise, and σ_t^2 is thermal noise. They are written as:

$$S = (DMP_r)^2 \quad (5.25)$$

$$\sigma_s^2 = 2eDM^{2+x} P_r B_e \quad (5.26)$$

$$\sigma_t^2 = \frac{4kTF_e}{R_L} B_e \quad (5.27)$$

$$D = \frac{\eta e}{h\nu} \quad (5.28)$$

where M is the APD multiplication factor, P_r is received optical power, e is electron charge, x is APD excess noise, B_e is receiver bandwidth, T is absolute temperature, F_e is the noise figure of the receiver, R_L is load resistance, η is the quantum efficiency of the detector, h is Planck's constant, and n is optical carrier frequency. $\sigma_{\Delta v}^2$ is noise due to LD linewidth, and σ_c^2 is equivalent noise due to interchannel crosstalk.

Equation 5.24 is transformed as:

$$\left(\frac{Q}{2}\right)^2 = \left\{ \left(\frac{S}{\sigma_s^2 + \sigma_t^2}\right)^{-1} + \left(\frac{S}{\sigma_{\Delta v}^2}\right)^{-1} + \left(\frac{S}{\sigma_c^2}\right)^{-1} \right\} \quad (5.29)$$

(a) Single channel transmission

In the case of single channel transmission, equation 5.29 is written as:

$$\left(\frac{Q}{2}\right)^2 = \left\{ \left(\frac{S}{\sigma_s^2 + \sigma_t^2}\right)^{-1} + \left(\frac{S}{\sigma_{\Delta v}^2}\right)^{-1} \right\} \quad (5.30)$$

LD linewidth dependence. Noise power due to the LD linewidth, Δv, is represented in Saito, Yamamoto and Kimura (1983) as:

$$\sigma_{\Delta v}^2 = K_i^2 \Delta v B_e \tag{5.31}$$

where K_i is the FM-to-AM conversion efficiency, and it is written as:

$$K_i = DMP_r \frac{\pi}{2f_d} \tag{5.32}$$

for an MZ-OFD at the center frequency. In equation 5.32, f_d is the frequency deviation of the FSK signal. From equations 5.30 to 5.32, the SNR determined by the LD linewidth is written as:

$$\text{SNR}_{\Delta v} = \frac{S}{\sigma_{\Delta v}^2} = \left(\frac{2}{\pi}\right)^2 \frac{f_d^2}{\Delta v B_e} \tag{5.33}$$

From equations 5.25, 5.26, 5.30 and 5.33 the power penalty at a BER of 10^{-9}, due to the LD linewidth, is written as:

$$\text{PP}_{\Delta v} = \frac{eM^x B_e + \left[(eM^x B_e)^2 + \left\{\left(\frac{2}{Q_0}\right) - \text{SNR}_{\Delta v}^{-1}\right\}\sigma_t^2\right]^{\frac{1}{2}}}{D\left\{\left(\frac{2}{Q_0}\right)^2 - \text{SNR}_{\Delta v}^{-1}\right\}P_{r0}} \tag{5.34}$$

where $Q_0 = 12$, and P_{r0} is received power at $\Delta v = 0$. The calculated power penalty at the BER of 10^{-9}, due to the LD linewidth, is shown in Fig. 5.33. In this calculation, P_{r0} is set to -39.8 dBm at the bit rate B of 622 Mbit/s, which is determined by the experimental result described in Toba et al. (1990). The power penalty increases with an increase in the LD linewidth, and increases rapidly for $f_d/B = 1.6$. Values of $\Delta v/B$ of 0.006 and 0.033 are required for an f_d/B of 1.6 and 3.2 in order to keep the power penalty less than 0.5 dB.

Detuning of the OFD. When the optical frequency of the transmitter does not coincide with the center frequency of the frequency discriminator, the signal power is denoted by:

$$S = \left\{DMP_r \cos\left(\pi \frac{f_0}{f_d}\right)\right\}^2 \tag{5.35}$$

where f_0 is the detuning of the MZ-OFD. The power penalty due to the offset frequency is calculated by replacing equation 5.25 by 5.35. The calculated penalty

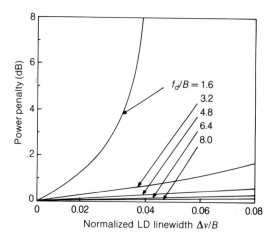

Fig. 5.33 Calculated power penalty due to LD linewidth for various frequency deviations.
Source: Toba, et al., 1991.

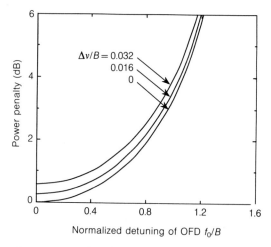

Fig. 5.34 Calculated power penalty due to detuning of the ORD for various LD linewidths.
Source: Toba, et al., 1991.

due to detuning of the OFD with LD linewidth as parameters is shown in Fig. 5.34. In this calculation, the normalized frequency deviation f_d/B is set at 3.2. For an LD linewidth $\Delta v/B$ of 0.016, the f_0/B should be less than 0.28 in order to keep the penalty less than 0.5 dB.

Frequency deviation discrepancy dependence. The discrepancy between the frequency deviation of the signal and frequency spacing of the MZ-OFD decreases the signal power, denoted as:

$$S = \left\{ DMP_r \sin\left(\pi \frac{f_m}{f_d}\right) \right\}^2 \tag{5.36}$$

where f_m is the *actual* frequency deviation of the FSK signal and f_d is the frequency spacing of the OFD, which is the *ideal* frequency deviation. The frequency deviation discrepancy characteristics are calculated by replacing equation 5.25 by 5.36. The calculated power penalty due to the frequency deviation discrepancy for the normalized *ideal* frequency deviation of $f_d/B = 3.2$ is shown in Fig. 5.35. Power penalty increases as the normalized frequency deviation changes more or less than 3.2, and it increases more rapidly at a smaller frequency deviation region for a larger linewidth. For $\Delta v/B = 0.016$, the normalized frequency deviation of the FSK signal should be $2.65 < f_m/B < 3.78$ for the power penalty to be less than 0.5 dB.

(b) Multichannel transmission

In the case of multichannel transmission, interchannel crosstalk should be taken into consideration. Cascaded MZ filters (periodic filters) are assumed to be used

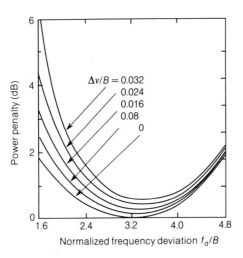

Fig. 5.35 • Calculated power penalty due to the frequency deviation discrepancy for the ideal normalized frequency deviation of $f_d/B = 3.2$.
Source: Toba, *et al.*, 1991.

for channel selection. The FSK spectrum in which the LD linewidth is taken into consideration is written as:

$$G_s(f) = \int G_{FSK}(f_0) \cdot G_p(f - f_0) df_0 \qquad (5.36a)$$

where $G_{FSK}(f)$ is the normalized FSK spectrum according to Lucky, Salz and Weldon (1968) and $G_p(f)$ is the carrier spectrum with:

$$G_p(f) = \frac{2\pi \Delta v}{\{2\pi(f - f_c)\}^2 + (\pi \Delta v)^2} \qquad (5.37)$$

The spectrum of interference channels is written as:

$$G_c(f) = \sum_{k=-N/2+1, \neq 0}^{N/2} G_s(f - k \cdot \Delta f) \quad (N: \text{even})$$

$$\sum_{k=-(N-1)/2, \neq 0}^{N-1/2} G_s(f - k \cdot \Delta f) \quad (N: \text{odd}) \qquad (5.38)$$

where N is the number of channels. The transmittance of the cascaded Mach-Zehnder filter is written as

$$T_{fil}(f) = \prod_{l=1}^{m} \frac{1}{2} \left\{ 1 + \cos\left(\pi \frac{f}{2^{l-1} \cdot \Delta f} \right) \right\} \qquad (5.39)$$

where m is the number of stages of Mach-Zehnder filters and N is equal to 2^m. For example, a seven-stage Mach-Zehnder filter can select one signal from among 128 multiplexed signals. The optical power of the signals is:

$$P_s = \int G_s(f) \cdot T_{fil}(f) df \qquad (5.40)$$

and that of the crosstalk is:

$$P_c = \int G_c(f) \cdot T_{fil}(f) df \qquad (5.41)$$

As a result, the equivalent signal-to-noise ratio determined by the crosstalk is:

$$SNR_c = \frac{S}{\sigma_c^2} = \left(\frac{P_s}{P_c} \right)^2 \qquad (5.42)$$

The power penalty at a BER of 10^{-9} is written as:

$$PP_c = \frac{eM^xB_e + \left[(eM^xB_e)^2 + \left\{\left(\frac{2}{Q_0}\right)^2 - SNR_{\Delta\nu}^{-1} - SNR_c^{-1}\right\}\sigma_t^2\right]^{\frac{1}{2}}}{D\left\{\left(\frac{2}{Q_0}\right)^2 - SNR_{\Delta\nu}^{-1} - SNR_c^{-1}\right\}P_{r0}} \quad (5.43)$$

On the basis of the above calculation, the following results are obtained.

Frequency spacing dependence. Frequency spacing dependence of the transmission characteristics for 128-channel signals with $f_d/B = 3.2$, and $\Delta\nu/B = 0.016$ are shown in Fig. 5.36. Crosstalk increases when frequency spacing decreases. The power penalty due to crosstalk also increases when frequency spacing decreases, and it increases steeply when normalized frequency spacing decreases to less than 10. Normalized frequency spacing should be more than 14 to keep the power penalty less than 0.5 dB.

The following calculation is carried out for $f_s/B = 16$ and $\Delta\nu/B = 0.016$ except where there are special comments. It corresponds to a frequency spacing of 10 GHz and linewidth of 10 MHz for a bit rate of 622 Mbit/s.

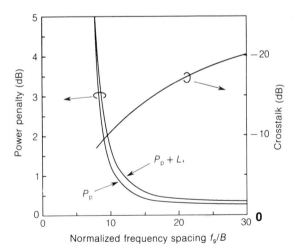

Fig. 5.36 Frequency spacing dependence of the transmission characteristics for 128-channel FSK signals at the normalized frequency deviation of 3.2 and the normalized linewidth of 0.016. P_p indicates the receiver sensitivity penalty and L_f indicates the transmittance loss of the FS-SW.
Source: Toba, et al., 1991.

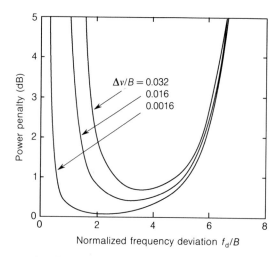

Fig. 5.37 Power penalty due to the frequency deviation for 128-channel FSK signals with the normalized frequency spacing of 16.
Source: Toba, et al., 1991.

Frequency deviation dependence. The receiver sensitivity degradation characteristics due to frequency deviation are shown in Fig. 5.37. In the calculation, the frequency spacing of the MZ-OFD is assumed to be equal to the frequency deviation of the FSK signals. For a $\Delta v/B$ of 0.016, the power penalty increases as f_d/B increases above 3.2, which is determined by the crosstalk characteristics. It also increases when f_d/B decreases to less than 3.2 due to the LD linewidth.

Power penalty degradation occurs more quickly as the LD linewidth becomes wider for a small frequency deviation range. As a result, there exists an optimum frequency deviation value for each LD linewidth, and the optimum value increases with the LD linewidth. For example, the minimum power penalty of 0.4 dB is obtained at $f_d/B = 3.2$ for $\Delta v/B = 0.016$, and the allowable normalized frequency deviation range from 2.64 to 4.10 for a power penalty of less than 0.5 dB.

Detuning of the FS-SW dependence. The frequency offset between the optical signals and the center frequency of the FS-SW increases the crosstalk penalty. Figure 5.38 shows the calculated results when the FS-SW detunes from the multiplexed signals. The frequency offset increases the crosstalk as well as the insertion loss. For example, the normalized offset frequency, f_{sw}/B, should be less than 0.8 to keep the total penalty increases to less than 0.1 dB, where the total penalty is 0.6 dB.

Signal power decrease dependence. Allowance for the output power variance of the channels is discussed in this section. This is important in constructing optical

OPTICAL FILTER AND DIRECT DETECTION

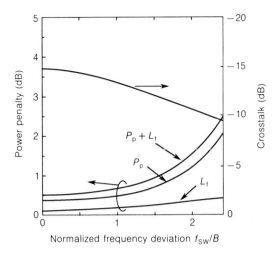

Fig. 5.38 Detuning of the frequency selection switch dependence of the transmission characteristics for 128-channel FSK signals with the normalized frequency spacing of 16, the normalized frequency deviation of 3.2 and normalized linewidth of 0.016.
Source: Toba, *et al.*, 1991.

FDM systems because it determines the requirement specification for the output power variance of the LDs and loss variance of the optical multiplexer and mixer. Figure 5.39 shows the calculated results of signal decrease dependence of the total penalty and crosstalk for 10 GHz-spaced, 128-channel FSK-direct detection transmission at bit rates of 156 Mbit/s and 622 Mbit/s; 156 Mbit/s and 622 Mbit/s are capable for transmission of NTSC and HDTV signals as converted by the linear CODEC.

In this calculation, the LD linewidth is 10 MHz, and the frequency deviation is 2 GHz. The basic receiver sensitivity, P_{r0}, is assumed to be -45.8 and -39.8 dBm for 156 Mbit/s and 622 Mbit/s, respectively. The interchannel crosstalk determines the penalty in this case. It depends on the inherent crosstalk value of the optical filter, C_t, which is denoted in Oda, *et al.* (1990) as:

$$C_t = 1 - (1 - C_f)^m \qquad (5.44)$$

where C_f is the inherent crosstalk of the one-stage MZ filter. The inherent crosstalk of the MZ filter degrades due to the unequal beam splitting ratio of the arms of the MZ filters, resultant polarization dependence, and imperfect control of the phase shifter. In this figure, the inherent crosstalk of the FS-SW is assumed to be -20 dB, which corresponds to the inherent crosstalk of a one-stage Mach-Zehnder filter of -28.4 dB. If the Mach-Zehnder filter has the

180 OPTICAL FREQUENCY DIVISION MULTIPLEXING SYSTEMS

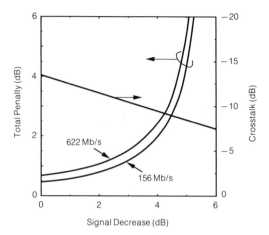

Fig. 5.39 Calculation results of the signal decrease dependence of the total penalty and crosstalk for 10 GHz-spaced, 128-channel transmission at bit rates of 156 Mbit/s and 622 Mbit/s.
Source: Toba, et al., 1991.

inherent crosstalk C_f per stage, then equation (5.39) is rewritten as

$$T_{fil}(f) = \prod_{l=1}^{m}(1-C_f)\left[\frac{1}{2}\left\{1+\cos\left(\pi\frac{f}{2^{l-1}\cdot \Delta f}\right)\right\}\right] \quad (5.39')$$

The calculation was performed using equation 5.39′ as a filter transmittance characteristic. The allowed signal decrease was -3.0 dB and 3.5 dB at 622 Mbit/s and 156 Mbit/s for a total penalty of 1.5 dB.

5.5.3 Transmission characteristics

A 10 GHz-spaced, 10-channel transmission experiment was performed by Toba, et al. (1991). Each output of the LD was multiplexed by a star coupler and one of the signals was selected by the waveguide type frequency selection switch (FSSW). Seven-stage MZ filters, whose frequency spacings were 10, 20, 40, 80, 160, 320 and 640 GHz, were used as the FSSW as in Toba, et al. (1990). Polarization-dependent transmittance characteristics are compensated for by the laser trimming the stress-applying films, which are loaded on the waveguide arms of the MZ filters according to Takato, et al. (1990). The selected FSK signals was discriminated and detected by the same scheme as in the single channel transmission.

OPTICAL FILTER AND DIRECT DETECTION

Multiplexed and frequency selected spectra for 622 Mbit/s FSK signals with a frequency deviation of 2 GHz are shown in Fig. 5.40. These were observed by using a scanning Fabry-Perot interferometer with an FSR of 5000 GHz and a finesse of 4000. Figure 5.40(a) shows the multiplexed FSK signals and (b) through

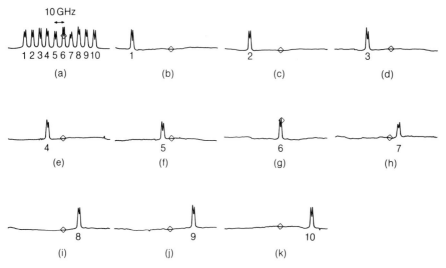

Fig. 5.40 Multiplexed and frequency selected spectra for 10 GHz-spaced, 622 Mbit/s FSK signals, with frequency deviation of 2 GHz. ◇ indicates the same frequency position in each spectrum. (a) Multiplexed spectrum; (b)–(k) channels 2 to 10 selected.
Source: Toba, *et al.*, 1991.

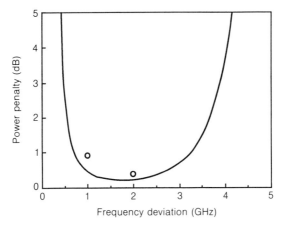

Fig. 5.41 Frequency deviation dependence of the power penalty for 10 GHz-spaced, 10-channel transmission at 622 Mbit/s. Solid line represents the calculated result for LD linewidth of 4 MHz.
Source: Toba, *et al.*, 1991.

(k) show the channel selected spectra. Frequency tuning was performed by adjusting the applied current for each thin-film heater of the Mach-Zehnder filter. The total applied power was 3.9 W on the average. Optical loss was about 7 dB, which was almost the same for each channel selection. The crosstalk component was hardly observed for each channel selected. The total crosstalk level was less than −13 dB. Receiver sensitivity degradation due to interchannel crosstalk hardly occurred.

The frequency deviation dependence of the receiver sensitivity degradation for 10-channel transmission is shown in Fig. 5.41. The solid line represents the calculated result for an LD linewidth of 4 MHz, which is based on Fig. 5.37. The experimental results almost agree with the calculated ones.

REFERENCES

Agrawal, G. P. (1987) Evaluation of crosstalk penalty in multichannel ASK heterodyne optical communication systems, *Electronics Letters*, **23**, pp. 906–908.

Al-Chalabi, S. A. *et al.* (1990) Temperature and mechanical vibration characteristics of a miniature long external cavity semiconductor laser, *Electronics Letters*, **26**, pp. 1159–1160.

Amann, M. C., Illek, S., Schanen, C. and Thulke, W. (1989b) Tuning range and threshold current of the tunable twin-guide (TTG) laser, *IEEE Photonics Technology Letters*, **1**, pp. 253–254.

Amann, M. C., Illek, S., Schanen, C., Thulke, W., and Lang, H. (1989a) Continuously tunable single-frequency laser diode utilizing transverse tuning scheme, *Electronics Letters*, **25**, pp. 837–839.

Bachus, E. J., Bohnke, F., Braun, R. P., Eutin, W., Foisel, H., Heimes, K. and Strebel, B. (1985a) Two channel heterodyne-type transmission experiments, *Electronic Letters*, **21**, pp. 35–36.

Bachus, E. J., Braun, R. P., Bohnke, F., Elze, G., Eutin, W., Foisel, H., Heimes, K. and Strebel, B. (1984) Digital transmission of TV signals with a fiber-optic heterodyne transmission system, *IEEE Journal Lightwave Technology*, **LT-2**, pp. 382–384.

Bachus, E. J., Braun, R. P., Caspar, C., Foisel, H. M., Grossmann, E., Strebel, B. and Westphal, F. J. (1989) Coherent optical multicarrier systems, *IEEE Journal Lightwave Technology*, **7**, pp. 375–384.

Bachus, E. J., Braun, R. P., Casper, C., Grossmann, E., Foisel, H., Heimes, K., Lamping, H., Strebel, B. and Westphal, F. J. (1986) Ten-channel coherent optical fiber transmission, *Electronics Letters*, **22**, pp. 1002–1003.

Bachus, E. J., Braun, R. P., Eutin, W., Grossmann, E., Foisel, H., Heimes, K. and Strebel, B. (1985b) Coherent optical fiber subscriber line, *IOOC-ECOC'85 Technical Digest*, **3**, pp. 61–64.

Baldacci, A., Ghersetti, S. and Rao, K. H. (1977) Interpretation of the acetylene spectrum at 1.5 μm, *Journal Molecular Spectroscopy*, **68**, pp. 183–194.

Barnes, J. A, *et al.* (1971) Characterization of frequency stability, *IEEE Transactions Instrumentation Measurement*, **IM-20**, pp. 105–120.

REFERENCES

Cheng, Y. H. and Okoshi, T. (1990) Multichannel DPFSK coherent optical communication systems, *Electronics Letters*, **26**, pp. 1378–1380.

Chikama, T., Naito, T., Watanabe, S., Kiyonaga, T., Suyama, M. and Kuwahara, H. (1990) Optical heterodyne image-rejection receiver for high-density optical frequency division multiplexing system, *IEEE Journal Selected Areas Communication*, **8**, pp. 1087–1094.

Chraplyvy, A. R. (1984) Optical power limits in multichannel wavelength-division-multiplexed systems due to stimulated Raman scattering, *Electronics Letters*, **20**, pp. 58–59.

Chraplyvy, A. R. (1990) Limitations on lightwave communications imposed by optical-fiber nonlinearities, *IEEE Journal Lightwave Technology*, **8**, pp. 1548–1557.

Chraplyvy, A. R., Tkach, R. W., Gnauck, A. H., Kasper, B. L. and Derosier, R. M. (1989) 8 Gbit/s FSK modulation of DFB lasers with optical demodulation, *Electronics Letters*, **25**, pp. 319–321.

Chung, Y. C. (1990) Frequency-locked 1.3 and 1.5 µm semiconductor lasers for lightwave systems applications, *IEEE Journal Lightwave Technology*, **8**, pp. 869–876.

Chung, Y. C. and Derosier, R. M. (1990) Frequency-locking of 1.5 µm InGaAsP lasers to an atomic krypton line without dithering the laser frequency, *IEEE Photonics Technology Letters*, **2**, pp. 435–437.

DeLange, O. E. (1970) Wide-band optical communication systems: part II, frequency-division-multiplexing, *Proceedings of the IEEE*, **58**, pp. 1683–1690.

Elrefaie, A., Maeda, M. W. and Guru, R. (1989) Impact of laser linewidth on optical channel spacing requirements for multichannel FSK and ASK systems, *IEEE Photonics Technology Letters*, **1**, pp. 88–90.

Flaarønning, N., Frorud, J. O., Sotom, M., Vendrome, G., Da Loura, G., Gabriagues, J. M., Jacquet, J., Leclerc, D. and Benoit, J. (1990) Multichannel FSK transmission at 565 Mb/s using tunable three-electrode DFB lasers, *Electronics Letters*, **26**, pp. 869–870.

Fujiwara, M., Shimosaka, N., Nishio, M., Suzuki, S., Yamazaki, S., Murata, S., and Kaede, K. (1990) A coherent photonic wavelength-division switching for broad-band networks, *IEEE Journal Lightwave Technology*, **8**, pp. 416–422.

Glance, B. S. (1986) An optical heterodyne mixer providing image-frequency rejection, *IEEE Journal Lightwave Technology*, **LT-4**, pp. 1722–1725.

Glance, B. S. and Scaramucci, O. (1990) High-performance dense FDM coherent optical networks, *IEEE Journal Selected Areas Communications*, **8**, pp. 1043–1047.

Glance, B., Fitzgerald, P. J., Pollock, K. J., Stone, J., Burrus, C. A. and Stulz, L. W. (1987) Frequency stabilization of FDM optical signals, *Electronics Letters*, **23**, pp. 274–276.

Glance, B., Pollock, K. J., Burrus, C. A., Kasper, B. L., Eisenstein, G., and Stulz, L. W. (1988a) WDM coherent optical star network, *IEEE Journal Lightwave Technology*, **6**, pp. 67–72.

Glance, B., Pollock, K., Burrus, C. A., Kasper, B. L., Eisenstein, G. and Stulz, L. W. (1987) Density-spaced WDM coherent optical star network, *Electronics Letters*, **23**, pp. 875–876.

Glance, B., Stone, J., Pollock, K. J., Fitzgerald, P. J., Burrus C. A. Jr., Kasper, B. L. and Stulz, L. W. (1988b) Densely spaced FDM coherent star network with optical signals confined to equally spaced frequencies, *IEEE Journal Lightwave Technology*, **LT-6**, pp. 1770–1781.

Gordon, J. P. and Mollenauer, L. F. (1991) Phase noise in photonic communications systems using linear amplifier noise, *IEEE Journal Lightwave Technology*, **9**, pp. 1351–1353.

Hall, J. L. and Lee, S. A. (1976) Interferometric real-time display of CW dye laser wavelength with sub-Doppler accuracy, *Applied Physics Letters*, **29**, pp. 367–369.

Hamaide, J. P. and Emphlit, Ph. (1990) Limitation in long haul IM/DD optical fiber systems caused by chromatic dispersion and nonlinear Kerr effects, *Electronics Letters*, **26**, pp. 1451–1453.

Hamaide, J. P., Emphlit, P., Prigent, L., Audouin, O. and Gabriagues, J. M. (1992) Effects of chromatic dispersion, Kerr nonlinearities and amplifier noise in long PSK optical fibre systems, *Electronics Letters*, **28**, pp. 44–46.

Hunkin, D. J., Hill, G. R. and Stallard, W. A. (1986) Frequency-locking of external cavity semiconductor lasers using an optical comb generator, *Electronics Letters*, **22**, pp. 388–390.

IEEE (1990) *Selected Areas in Communications*, **8** (6), 1005–14.

Inoue, K. (1990) Wavelength conversion for frequency-modulated light using optical modulation to oscillation frequency of a DFB laser diode, *IEEE Journal Lightwave Technology*, **8**, pp. 906–911.

Inoue, K. and Toba, H. (1991) Error rate degradation due to fibre four-wave mixing in 4-channel FSK direct-detection transmission, *IEEE Photonics Technology Letters*, **3**, pp. 77–79.

Inoue, K. and Toba, H. (1992a) Theoretical evaluation of error rate degradation due to fiber four-wave mixing in multichannel FSK heterodyne envelope detection transmissions, *IEEE Journal Lightwave Technology*, **10**, pp. 361–366.

Inoue, K. and Toba, H. (1992b) Wavelength conversion experiment using fiber four-wave mixing, *IEEE Photonics Technology Letters*, **4**, pp. 69–72.

Inoue, K., Toba, H. and Nosu, K. (1991) Multichannel amplification utilizing an Er^{3+}-doped fiber amplifier, *IEEE Journal Lightwave Technology*, **9**, pp. 368–374.

Inoue, K., Toba, H. and Oda, K. (1992) Influence of fiber four-wave mixing on multichannel FSK direct detection systems, *IEEE Journal Lightwave Technology*, **10**, pp. 350–360.

Ishida, O. (1991) Lightwave frequency tracking with a tunable DBR laser, *IEEE Journal Lightwave Technology*, **9**, pp. 1083–1093.

Ishida, O. and Toba, H. (1991) Lightwave frequency synthesizer with lock-in-detected frequency references, *IEEE Journal Lightwave Technology*, **9**, pp. 1344–1352.

Ishida, O. and Toba, H. (1991a) Submegahertz absolute frequency stability in a 1.5 µm tunable diode laser, *CLEO'91 Technical Digest*, **CThB2**, Baltimore.

Ishida, O. and Toba, H. (1991b) 200 kHz absolute frequency stability in a 1.5 µm external-cavity semiconductor laser, *Electronics Letters*, **27**, pp. 1018–1019.

Ishikawa, J., Ito, N. and Tanaka, K. (1986) Accurate wavelength meter for CW lasers, *Applied Optics*, **25**, pp. 639–643.

Iwashita, K. and Takatio, N. (1990) Chromatic dispersion compensation in coherent optical communications, *IEEE Journal Lightwave Technology*, **8**, 367–375.

Jennings, D. A., Evenson, K. M. and Knight, D. J. E. (1986) Optical frequency measurements, *Proceedings IEEE*, **74**, pp. 168–179.

Kaminow, I. P. (1990) FSK with direct detection in optical multiple-access FDM networks, *IEEE Journal Selected Areas Communications*, **8**, pp. 1005–1014.

Kaminow, I. P., Iannone, P. P., Stone, J. and Stulz, L. W. (1988) FDMA-FSK star network with a tunable optical filter demultiplexer, *IEEE Journal Lightwave Technology*, **6**, pp. 1406–1414.

Kaminow, I. P., Iannone, P. P., Stone, J. and Stulz, L. W. (1987) FDM-FSK star network with a tunable optical filter demultiplexer, *Electronics Letters*, **23**, pp. 1102–1103.

Kartaschoff, P. (1978) *Frequency and Time*, Section 2, Academic Press.

Kazovsky, L. G. (1987a) Multichannel coherent optical communications systems, *IEEE Journal Lightwave Technology*, **LT-5**, pp. 1095–1102.

Kazovsky, L. G. (1987b) Multichannel coherent optical communication systems, *OFC/IOOC'87 Technical Digest*, **TUG1**.

Kazovsky, L. G. and Gimlett, J. L. (1988) Sensitivity penalty in multichannel coherent optical communications, *IEEE Journal Lightwave Technology*, **6**, pp. 1353–1365.

Kazovsky, L. G. and Jacobsen, G. (1989) Multichannel CPFSK coherent optical communications systems, *IEEE Journal Lightwave Technology*, **7**, pp. 972–982.

Kobayashi, K. and Mito, I. (1988) Single frequency and tunable laser diodes, *IEEE Journal Lightwave Technology*, **6**, pp. 1623–1633.

Kotaki, K., Matsuda, M., Yano, M., Ishikawa, H. and Imai, H. (1987) 1.55 µm wavelength tunable FBH-DBR laser, *Electronics Letters*, **23**, pp. 325–327.

Kotaki, Y., Ogita, S., Matsuda, M., Kuwahara, Y. and Ishikawa, H. (1989) Tunable narrow-linewidth and high-power 2/4-shifted DFB laser, *Electronics Letters*, **25**, pp. 990–992.

Kourogi, K., Nakagawa, K., Shin, C. H., Teshima, M. and Ohtsu, M. (1991) Accurate frequency measurement system for 1.5 µm wavelength laser diodes, *CLEO'91 Technical Digest*, **CThR 57**, Baltimore.

Kuboki, K., Sasaki, S., Kitajima, S., Tsushima, H. and Yamashita, K. (1990) A channel-spacing stabilization using spectrum-slice synchronous-detection method for coherent FDM distributing system, in *Proceedings OEC'90*, **12C3-4**, pp. 152–153.

Kuznetsov, M. (1988) Theory of wavelength tuning in two-segment distributed feedback lasers, *IEEE Journal Quantum Electronics*, **24**, pp. 1837–1844.

Lucky, R. W., Salz, J. and Weldon, E. J. Jr. (1968) *Principles of Data Communication*, McGraw-Hill, New York.

Maeda, M. W. and Kazovsky, L. G. (1989) Novel frequency stabilization technique for multichannel optical communication systems, *IEEE Photonics Technology Letters*, **1**, pp. 455–457.

Maeda, M. W., Barry, J. R., Kumazawa, T. and Wagner, R. E. (1989) Absolute frequency identification and stabilization of DFB lasers in 1.5 µm region, *Electronics Letters*, **25**, pp. 9–11.

Maeda, M. W. et al. (1990) The effect of four-wave mixing in fibers on optical frequency-division multiplexed systems, *IEEE Journal Lightwave Technology*, **8**, pp. 1402–1408.

Malyon, D. and Stallard, W. A. (1989) Crosstalk in optical amplifiers and low chirp laser, *Electronics Letters*, **25**, pp. 495–496.

Marcuse, D. (1991) Single-channel operation in very long nonlinear fibers with optical amplifiers at zero dispersion, *IEEE Journal Lightwave Technology*, **9**, pp. 356–361.

Metrologia (1984) Documents concerning the new definition of the metre, **19**, pp. 163–177.

Murata, S., Mito, I. and Kobayashi, K. (1987) Over 720 GHz (5.8 nm) frequency tuning by a 1.5 µm DBR laser with phase and Bragg wavelength control section, *Electronics Letters*, **23**, pp. 403–405.

Mutrata, S., Mito, I. and Kobayashi, K. (1988) Tuning ranges for 1.5 µm wavelength tunable DBR lasers, *Electronics Letters*, **24**, pp. 577–578.

Naito, T., Chikama, T., Suyama, M., Kiyonga, T. and Kuwahara, H. (1989) Crosstalk penalty in a two-channel 560-Mbit/s DPSK heterodyne optical communication system using an image rejection receiver, *OFC'89 Technical Digest*, p. 141.

Noé, R., Rodler, H., Gaukel, G., Noll, B. and Ebberg, A. (1990) High-performance, two-channel optical FSK heterodyne system with polarization diversity receiver, *Electronics Letters*, **26**, pp. 1109–1110.

Norimatsu, S. and Iwashita, K. (1991) Cross-phase modulation influence on a two-channel optical PSK homodyne transmission system, *IEEE Photonics Technology Letters*, **3**, pp. 1142–1144.

Nosu, K., Toba, H. and Iwashita, K. (1987) Optical FDM transmission, *IEEE Journal Lightwave Technology*, **LT-5**, pp. 1301–1308.

Öberg, M., Nilsson, S., Klinga, T. and Ojala, P. (1991) A three-electrode distributed Bragg reflector laser with 22 nm wavelength tuning range, *IEEE Photonics Technology Letters*, **3**, pp. 299–301.

Oda, K., Takato, N., Kominato, T. and Toba, H. (1990) A 16-channel frequency selection switch for optical FDM distribution systems, *IEEE Journal Selected Areas Communications*, **8**, pp. 1132–1140.

Ohtsu, M. (1988) Realization of ultrahigh coherence in semiconductor lasers by negative electrical feedback, *IEEE Journal Lightwave Technology*, **6**, pp. 245–256.

Okoshi, T., Emura, K., Kikuchi, K. and Kersten, T. R. (1981) Computation of bit-error rate of various heterodyne and coherent-type optical communication schemes, *Journal Optical Communications*, **2**, pp. 89–96.

Olsson, N. A., Hegarty, J., Logan, R. A., Johnson, L. F., Walker, K. L., Cohen, L. G., Kasper, B. L. and Campbell, J. C. (1985) 68.3 km transmission wtih 1.37 Tbit km/s capacity using wavelength division multiplexing of ten single-frequency lasers at 1.5 μm, *Electronics Letters*, **21**, pp. 105–106.

Park, Y. K., Bergstein, S. S., Tench, R. E., Smith, R. W., Korotky, S. K., Burns, K. J. and Granlund, S. W. (1988) Crosstalk and prefiltering in a two-channel ASK heterodyne detection system without the effect of laser phase noise, *IEEE Journal Lightwave Technology*, **6**, pp. 1312–1320.

Saito, S., Murakami, M., Naka, A., Fukada, Y., Imai, T., Aiki, M. and Ito, T. (1991) 2.5 Gbit/s, 80–100 km spaced in-line amplifier transmission experiments over 2500–4500 km, *Proceedings ECOC'91*, Paris, France, post-deadline paper A.PDP.5.

Saito, S., Yamamoto, Y. and Kimura, T. (1982) Semiconductor laser FSK modulation and optical direct discriminator detection, *Electronics Letters*, **18**, pp. 468–469.

Saito, S., Yamamoto, Y. and Kimura, T. (1983) S/N and error rate evaluation for an optical FSK-heterodyne detection system using semiconductor lasers, *IEEE Journal Quantum Electronics*, **QE-19**, pp. 180–193.

Sakai, Y., Kano, F. and Sudo, S. (1990) Small-frequency-difference stabilization of laser diodes using $^{12}C_2H_2$ and $^{13}C_2H_2$ absorption lines for transmitter and local oscillator of optical heterodyne systems, *IEEE Photonics Technology Letters*, **2**, pp. 762–765.

Sasada, H. and Yamada, K. (1990) Calibration lines of HCN in the 1.5 μm region, *Applied Optics*, **29**, pp. 3535–3547.

Shibata, N., Braun, R. P. and Waarts, R. G. (1986) Crosstalk due to three-wave mixing process in a coherent single-mode transmission line, *Electronics Letters*, **22**, pp. 675–677.

Shibata, N., Braun, R. P. and Waarts, R. G. (1987) Phase-mismatch dependence of efficiency of wave generation through four-wave mixing in a single-mode optical fiber, *IEEE Journal Quantum Electronics*, **QE-23**, pp. 1205–1210.

Shibata, N., Iwashita, K. and Azuma, Y. (1989) Receiver sensitivity degradation due to four-wave mixing in a 2 Gb/s CPFSK heterodyne transmission system, in *IOOC'89 Technical Digest*, Kobe, Japan, paper 18C1-3.

Shibata, N., Nosu, K., Iwashita, K. and Azuma, Y. (1990) Transmission limitations due to fiber nonlinearities in optical FDM systems, *IEEE Journal Selected Areas Communications*, **8**, pp. 1068–1077.

REFERENCES

Shibutani, M., Yamazaki, S., Shimosaka, N., Murata, S. and Shikada, M. (1989) Ten-channel coherent optical FDM broadcasting system, *OFC'89 Technical Digest*, **THC2**, Houston, TX, p. 140.

Shimosaka, N., Kaede, K., Fujiwara, M., Yamazaki, S., Murata, S. and Nishio, M. (1990) Frequency separation locking and synchronization for FDM optical sources using widely frequency tunable laser diodes, *IEEE Journal Selected Areas Communications*, **8**, pp. 1078–1086.

Shirasaki, M., Yokota, I. and Touge, T. (1990) 20 Gbit/s no-chirp intensity modulation by DPSH-IM method and its fiber transmission through 300 ps/nm dispersion, *Electronics Letters*, **26**, pp. 33–35.

Stanley, I. W., Hill, G. R. and Smith, D. W. (1987) The application of coherent optical techniques to wide-band networks, *IEEE Journal Lightwave Technology*, **LT-5**, pp. 439–451.

Suyama, M., Chikama, T. and Kuwahara, H. (1988) Channel allocation and crosstalk penalty in coherent optical frequency division multiplexing systems, *Electronics Letters*, **24**, pp. 1278–1279.

Takato, N., Sugita, A., Onose, K., Okazaki, H., Okuno, M., Kawachi, M. and Oda, K. (1990) 128-channel polarization-insensitive frequency-selection-switch using high-silica waveguides on Si, *IEEE Photonics Technology Letters*, **2**, pp. 441–443.

Telle, H. R., Meschede, D. and Hänsch, T. W. (1990) Realization of a new concept for visible frequency division: phase locking of harmonic and sum frequencies, *Optics Letters*, **15**, pp. 532–534.

Toba, H., Inoue, K. and Nosu, K. (1986) A conceptional design on optical frequency-division-multiplexing distribution systems with optical tunable filters, *IEEE Journal Selected Areas Communications*, **22**, pp. 1458–1467.

Toba, H., Inoue, K., Nosu, K. and Motosugi, G. (1988) A multichannel laser diode frequency stabilizer for narrowly spaced optical frequency-division-multiplexing transmission, *Journal Optical Communications*, **9**, pp. 3–7.

Toba, H., Oda, K., Nakanishi, K., Shibata, N., Nosu, K., Takato, N. and Fukuda, M. (1990) A 100-channel optical FDM transmission/distribution at 622 Mb/s over 50 km, *IEEE Journal Lightwave Technology*, **8**, pp. 1396–1401.

Toba, H., Oda, K. and Nosu, K. (1989) A 16-channel optical FDM distribution/transmission experiment, *IOOC'89 Technical Digest*, pp. 174–175.

Toba, H., Oda, K. and Nosu, K. (1991) Design and performance of FSK-direct detection scheme for optical FDM systems, *IEEE Journal Lightwave Technology*, **9**, pp. 1335–1343.

Toba, H., Oda, K., Nosu, K., Takato, N. and Miyazawa, H. (1988) 5 GHz-spaced eight-channel optical FDM transmission experiment using guided-wave tunable demultiplexer, *Electronics Letters*, **24**, pp. 78–80.

Tsushima, H., Sasaki, S., Kuboki, K., Kitajima, S., Takeyari, R. and Okai, M. (1991) 1.244 Gbit/s 32-channel 121 km transmission experiment using shelf-mounted CPFSK optical heterodyne system, in *Proceedings ECOC'91/IOOC'91*, Paris, France, pp. 397–400.

Vodhanel, R. S., Elregaie, A. E., Iqbal, M. Z., Wagner, R. E., Gimlett, J. L. and Tsuji, S. (1990) Performance of directly modulated DFB lasers in 10 Gb/s ASK, FSK, and DPSK lightwave systems. *IEEE Journal Lightwave Technology*, **8**, pp. 1379–1386.

Waarts, R. G. and Braun, R. P. (1985) Crosstalk due to stimulated Brillouin scattering in monomode fibre, *Electronics Letters*, **21**, pp. 1114–1115.

Waarts, R. G. and Braun, R. P. (1986) System limitations due to four-wave mixing in single-mode optical fibre, *Electronics Letters*, **22**, pp. 873–874.

Welter, R., Sessa, W. B., Maeda, M. W., Wagner, R. E., Curtis, L., Young, J., Lee, T. P., Nanduri, K., Kodera, H., Koga, Y. and Barry, J. R. (1989) Sixteen-channel coherent broadcast network at 155 Mbit/s, *IEEE Journal Lightwave Technology*, **7**, pp. 1438–1444.

Willner, A. E. (1990) Simplified model of an FSK-to-ASK direct-detection system using a Fabry-Perot demodulator, *IEEE Photonics Technology Letters*, **2**, pp. 363–366.

Willner, A. E., Kaminow, I. P., Kuznetsov, M., Stone, J. and Stulz, L. W. (1990) 1.2 Gb/s closely-spaced FDMA-FSK direct-detection star network, *IEEE Photonics Technology Letters*, **2**, pp. 223–226.

Wuenstel, K., Schweizer, H., Schiling, M., Idler, W., Kuehn, E., Laube, G. and Hildebrand, O. (1990) Integrated interferometric injection lasers (L = 1300 nm and 1500 nm) with tuning range exceeding 20 nm, *12th IEEE International Laser Conference, Technical Digest*, Davos, Switzerland, pp. 212–213.

Wyatt, R. and Delvin, W. J. (1983) 10 kHz linewidth 1.5 µm InGaAsP external cavity lasers with 55 nm tuning range, *Electronics Letters*, **19**, pp. 110–112.

Yamazaki, S., Ono, T., Shimizu, H. and Emura, K. (1990a) A bidirectional common polarization control method for coherent optical FDM transmission system, *IEEE Photonics Technology Letters*, **2**, pp. 135–138.

Yamazaki, S., Ono, T., Shimizu, H., Kitamura, M. and Emura, K. (1990b) 2.5 Gb/s CPFSK coherent multichannel transmission experiment for high capacity trunk line system, *IEEE Photonics Technology Letters*, **2**, pp. 914–916.

Yamazaki, S., Shibutani, M., Shimosaka, N., Murata, S., Ono, T., Kitumara, M., Emura, K. and Shikada, M. (1990c) A coherent optical FDM CATV distribution system, *IEEE Journal Lightwave Technology*, **8**, pp. 396–405.

Yanagawa, T., Saito, S., Machida, S. and Yamamoto, Y. (1985) Frequency stabilization of an InGaAsP distributed feedback laser to an NH_3 absorption line at 15137 Å with an external frequency modulator, *Applied Physics Letters*, **47/10**, pp. 1036–1038.

Yasaka, H. and Kawaguchi, H. (1988) Linewidth reduction and optical frequency stabilization of a distributed feedback laser by incoherent optical negative feedback, *Applied Physics Letters*, **53**, pp. 1360–1362.

Yoshikuni, Y., Oe, K., Motosugi, G. and Matsuoka, T. (1986) Broad wavelength tuning under single-mode oscillation with a multi-electrode distributed feedback laser, *Electronics Letters*, **22**, pp. 1153–1154.

6
Optical amplifiers for coherent transmission and optical FDM

Optical amplifiers have been studied for a long time, starting with optical amplification experiments carried out in 1960. At that time, the results did not confirm their practical application in the communications field. Around 1970, the first continuous oscillation of laser diodes (LDs) at room temperature was achieved. The initial development of optical devices as light sources in communications focused on LDs. From the mid 1980s, research into optical fiber amplifiers became active again. In 1985, low-loss optical fibers doped with rare-earth ions were developed, and in 1987, optical amplification using an erbium-doped single-mode fiber was successfully carried out. Furthermore, in 1989, successful transmission experiments were reported. Through those transmission experiments, optical amplifiers were designated a new technology, leading to a new era in optical communications.

Combined with coherent technologies, optical amplifiers have excellent capabilities as in-line amplifiers for long span transmission, or as common amplifiers for high density wavelength division multiplexing. In this chapter, we focus on two applications: in-line amplifier systems and common amplifier systems. Their system performances are discussed theoretically and experimentally, and the guidelines of system design are clarified.

6.1 IN-LINE AMPLIFIER SYSTEMS

Takeshi Ito

Optical fiber amplifiers possess inherent high matching properties such as easy coupling with the optical fiber used for transmission and little polarization dependence with respect to amplification characteristics. Cascaded fiber amplifiers are, therefore, expected to be used in long span transmission systems. System limitations of cascaded fiber amplifiers are discussed in this section. First, the theoretical bounds are discussed, and the latest transmission performance achieved with cascaded erbium-doped fiber amplifier systems is given.

6.1.1 System performance and amplifier performance

The performance of any cascaded optical amplifier system can be evaluated by modifying the equations proposed by Olsson (1989). His work addressed the noise accumulation in cascaded optical amplifier systems. The modified equation given in Ito (1991) is:

$$B \cdot L = \frac{1}{\text{SNR} \cdot \frac{B_e}{B}} \frac{l}{e^{\alpha l} - 1} \frac{1}{h\nu} \frac{P_s}{\text{NF}} \quad (6.1)$$

for heterodyne detection, and

$$B \cdot L = \frac{2}{\text{SNR} \cdot \frac{B_e}{B}} \frac{l}{e^{\alpha l} - 1} \frac{1}{h\nu} \frac{P_s}{\text{NF}} \quad (6.2)$$

for homodyne detection, where:

B = transmission bit rate;
L = transmission distance;
SNR = received signal-to-noise ratio;
B_e = electrical bandwidth;
α = fiber attenuation constant;
l = in-line amplifier spacing;
$h\nu$ = photon energy;
P_s = in-line amplifier output power;
NF = in-line amplifier noise figure.

A schematic model of the cascaded amplifier system is shown in Fig. 6.1. In the model, in-line amplifiers with the same output power P_s, noise figure NF and optical gain G are placed at equal intervals. Each compensates for the optical signal loss occurring in each fiber section. Therefore, $e^{\alpha l}$ in equations 6.1 and 6.2 equals G.

In cascaded amplifier systems, signal spontaneous beat noise, spontaneous/spontaneous beat noise, local oscillator/spontaneous beat noise, and local oscillator shot noise are present in addition to receiver thermal noise. If many cascaded amplifiers with high optical gain are used, local oscillator spontaneous beat noise predominates. In such cases, equations 6.1 and 6.2 are suitable for assessing system performance.

The left side of both equations is the product of transmission bit rate, B, and transmission distance, L. The first term of the right side is a function of the received signal-to-noise ratio required to achieve a specified transmission quality. The term depends on the detection scheme used. The product of B and L decreases

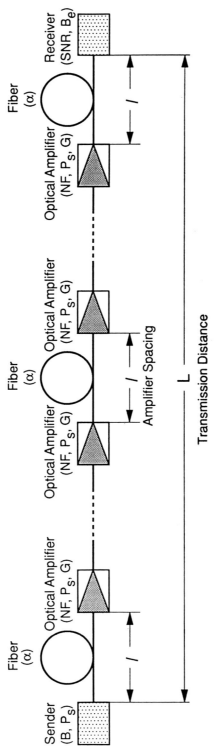

Fig. 6.1 Cascaded amplifier system model.

as the required SNR increases for both schemes. The second term is a function of amplifier spacing, l. The last term reflects the optical amplifier performance as defined by the ratio of the optical amplifier output, P_s, and the optical amplifier noise figure, NF.

6.1.2 Limitation factors

System performance, as defined by the product of B and L, increases as the amplifier spacing decreases. This is because the optical gain of each amplifier can be decreased as the amplifier spacing decreases, and optical amplifier noise is proportional to the gain. As a rough example, if the original spacing is 100 km and is reduced by half, the optical gain can be reduced by 12.5 dB, assuming a fiber loss of 0.25 dB/km. The lower total fiber loss due to the decreased spacing means that the optical gain can be decreased. Theoretically we would expect the accumulated noise to be also decreased by 12.5 dB. Unfortunately, the total number of optical amplifiers required to achieve the same transmission distance must be doubled, which increases the noise by at most 3 dB. As a result, the accumulated noise is decreased by 9.5 dB. Therefore, the product of B and L increases as the amplifier spacing is decreased.

In terms of amplifier spacing, system performance is limited by noise accumulation, which is shown by line (i) in Fig. 6.2. The product of B and L also increases

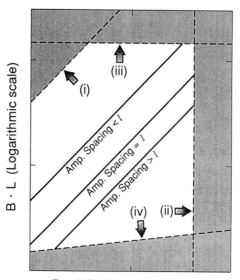

Fig. 6.2 Limitation factors to system performance.

with improved amplifier performance; defined here as P_s/NF. The system performance is limited by amplifier performance, as shown by line (ii). In addition, the product of B and L is also limited by fiber dispersion, as shown by line (iii). Needless to say, cascaded systems are not needed in the region of small BL values that lie under line (iv).

Nonlinear effects are a serious problem, especially for cascaded systems that use phase detection schemes. Performance limits due to nonlinear effects are not shown in this figure because the relationship between transmission distance and transmission bit rate is complicated. This subject is discussed later on.

In order to achieve the required system performance within the allowable region, the relationship between amplifier spacing and amplifier performance should be optimized.

6.1.3 Fiber dispersion limit

Waveform distortion due to fiber dispersion is a serious problem in cascaded amplifier systems because the optical signal is transmitted without regeneration over long distances. The fiber dispersion limit is given by:

$$B^2 \cdot L = \gamma \cdot \frac{c}{D(\lambda) \cdot \lambda^2} \tag{6.3}$$

where γ = fiber dispersion limit coefficient, c = light velocity, λ = wavelength, $D(\lambda)$ = fiber dispersion.

The product of the square of the transmission bit rate and the transmission distance is limited by fiber dispersion. The fiber dispersion limit coefficient, γ, depends on the modulation/detection scheme used, and ranges from 0.3 to 0.74 for CPFSK with a modulation index of 1 and DPSK, respectively, as in Elrefaie, et al. (1988) and Iwashita and Takachio (1990). The criterion of fiber dispersion limit is a power penalty of 1 dB caused by fiber dispersion. The fiber dispersion limit of CPFSK systems is the most severe. However, the limit can be increased by using delay equalization in the intermediate frequency band together with heterodyne detection. As an additional remark, the fiber dispersion limit of the IM/DD systems is the least severe, and the coefficient, γ, is 0.79.

Based on coefficient γ, the fiber dispersion limits of IM/DD and CPFSK/heterodyne systems are shown in Fig. 6.3. All discrete values are experimental data. The limit of a CPFSK/heterodyne system using intermediate frequency delay equalization is also shown. The figure assumes a fiber dispersion of 0.5 ps/nm/km. In addition, the Gordon-Haus limit discussed in Gordon and Haus (1986) of soliton transmission is also shown. Below approximately 10 Gbit/s, the CPFSK/heterodyne system with intermediate frequency delay equalization is superior to

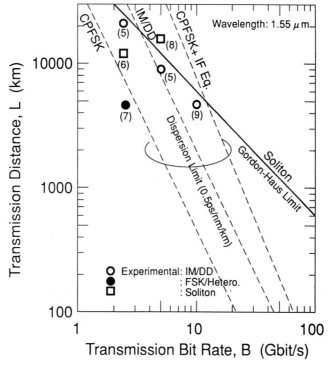

Fig. 6.3 Transmission distance limited by the fiber dispersion.
References:
(5) Bergaro, et al. (1991)
(6) Mollenauer, et al. (1991)
(7) Saito, et al. (1991)
(8) Mollenauer, et al. (1992)
(9) Taga, et al. (1992)

an optical soliton system from the dispersion limit viewpoint. The experimental data closely approaches its respective limits.

6.1.4 Noise accumulation limit

As the amplifier spacing is decreased, the accumulated noise decreases but more amplifiers must be cascaded. Roughly speaking, the maximum BL values is achieved when the spacing is 0 km. Obviously this is not realistic but it does serve as a useful reference. This is shown by the solid line 1 in Fig. 6.4 for the CPFSK/heterodyne system. The required SNR is assumed to be a level of achieving an error rate of 10^{-11}. Other parameters assumed are $B_e/B = 1.6$, $\alpha = 0.21$ dB/km, $\lambda = 1.552$ μm ($v \fallingdotseq 193.3$ THz), $\gamma = 0.47$ and $D(\lambda) = 0.5$ ps/nm/km.

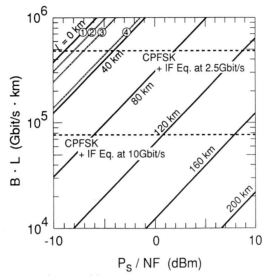

Fig. 6.4 Noise accumulation limit. Key:
$\frac{v}{B} = 0$, $\frac{v}{B} = 5 \times 10^{-4}$, $\frac{v}{B} = 10^{-3}$, $\frac{v}{B} = 5 \times 10^{-3}$

In the heterodyne system, B_e means a bandwidth of the intermediate frequency band. The horizontal lines indicate the dispersion limits of the CPFSK/heterodyne system with intermediate frequency band equalization at 2.5 and at 10 Gbit/s. If the amplifier spacing is decreased by increasing P_s/NF, the product of B and L can reach the dispersion limit, neglecting noise accumulation.

Noise accumulation increases with nonideal device performance. The broad spectrum, $\Delta v/B$, of the signal and local lasers lowers the BL value in the CPFSK/heterodyne system. For example, when $\Delta v/B = 10^{-3}$, BL is about 70% its value when $\Delta v/B = 0$, as shown by solid or dotted lines 1 and 3. However, the noise accumulation limit even in such cases yields a system performance that exceeds the maximum required system performance of $BL = 10^5$ for a 10 Gbit/s transoceanic transmission system. A broader spectrum of the signal and local lasers compared with the ideal laser spectrum impacts very little on the amplifier spacing.

BL is also limited by the optical amplifier performance defined by P_s/NF. The state of the art in EDFA performance is shown in Fig. 6.5. The open circles show the performance of the amplifiers used in the 2200 km transmission experiment as in Saito, Imai and Ito (1991). NF increases in the high output power region. The maximum P_s/NF was 0 dBm. The EDF had an erbium concentration of 96 ppm, and a length of 160 m. A P_s/NF value of 10 dBm was recorded in a 48 m long, center codoped fiber with an erbium concentration of 1000 ppm and aluminum concentration of 5000 ppm. No degradation in noise figure was found

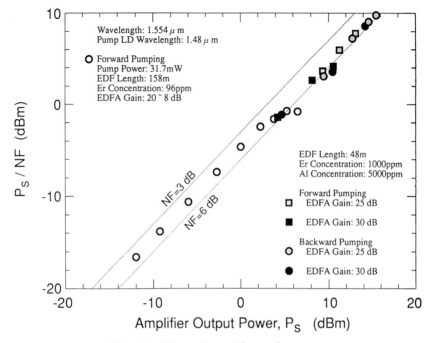

Fig. 6.5 Observed amplifier performance.

even in the high output power region around 15 dBm. Measured P_s/NF values did not depend on the pumping schemes, as shown in Fig. 6.5.

Optical amplifiers are constructed with optical components such as couplers, isolators, filters, and so on. In addition, optical components for supervisory and automatic power control (APC) circuits are needed to realize a practical system. The loss of optical components inserted into the input end of an EDFA increases the net noise figure. The loss of optical components inserted into the output end of an EDFA decreases the net output power. As a result, the insertion losses of these components degrade the net P_s/NF value. Amplifier performance should be evaluated by the overall P_s/NF, not by the net P_s/NF.

6.1.5 Nonlinear effect limit

In cascaded amplifier systems, nonlinear effects appear as the result of long transmission distances even if the amplifier output power is not high. Through the Kerr effect, intensity fluctuations in optical signals cause phase fluctuations in the optical carrier wave. This phase fluctuation is a serious problem in a coherent detection system, and limits the attainable *BL* product.

IN-LINE AMPLIFIER SYSTEMS

There are three plausible origins of intensity changes in coherent systems. These are:

(i) residual intensity modulation accompanying phase (frequency) modulation;
(ii) FM-to-AM conversion stemming from fiber dispersion;
(iii) intensity fluctuation resulting from amplified spontaneous emission (ASE) noise accumulation in cascaded amplifiers.

Effects from the first two origins can be reduced sufficiently by precisely controlling modulation conditions or minimizing fiber dispersion. The third is inherent to cascaded amplifier systems, and its effect cannot always be neglected.

The limit due to the carrier phase fluctuations caused by ASE through the Kerr effect is given by Gordon and Mollenauer (1990) as:

$$B^3 \cdot L = 3\sigma_\phi^2 \cdot \left(\frac{\alpha}{k_2}\right)^2 \cdot l^3 \cdot \frac{1}{e^{\alpha l} - 1} \cdot \frac{1}{h\nu} \cdot \frac{1}{P_s \cdot NF} \quad (6.4)$$

where σ_ϕ^2 = phase noise allowable to attain a predetermined error rate, k_2 = nonlinear coefficient given by:

$$k_2 = \frac{2\pi n_2}{\lambda \cdot A_{eff}} \quad (6.5)$$

where n_2 = Kerr coefficient, A_{eff} = effective fiber core area.

The product of transmission bit rate and the cube of the transmission distance is inversely proportional to the Kerr constant squared, amplifier output power and amplifier noise figure. The phase noise σ_ϕ^2 depends on the detection scheme used. With PSK heterodyne synchronous detection or PSK homodyne detection, σ_ϕ^2 must be less than 0.053 to achieve an error rate of less than 10^{-11}. With CPFSK heterodyne-delay detection or DPSK heterodyne detection, the corresponding σ_ϕ^2 value must be less than 0.027.

The product BL^3 for CPFSK heterodyne-delay detection as a function of amplifier output power is shown in Fig. 6.6. With $l = 40$ km, which is almost the optimum spacing from the viewpoint of the nonlinear effects, an amplifier output power of -10 dBm yields an expected BL^3 of 8×10^{12} Gbit/s·km³, which corresponds to an 8 Gbit/s transoceanic system.

6.1.6 Expected system performance

Ultimate system performance is basically estimated with the limits mentioned above. The overall performance given in Saito, Aiki and Ito (1992) of cascaded

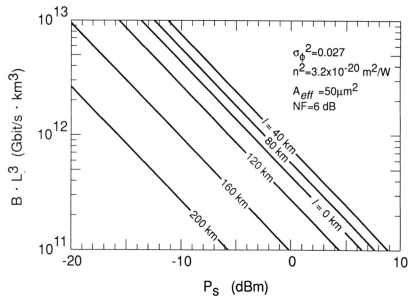

Fig. 6.6 Nonlinear effect limit.

amplifier systems using CPFSK modulation and heterodyne-delay detection is shown in Fig. 6.7. The horizontal broken line shows the fiber dispersion limit with delay equalization in the intermediate frequency band at 10 Gbit/s. In addition, both the noise accumulation limit and nonlinear effect limit must be considered. Under the noise accumulation limit condition, an attainable transmission distance, L_{noise}, decreases as the amplifier spacing increases at a specific amplifier output power, and increases as the amplifier output power increases at a specific amplifier spacing, as shown in Fig. 6.4. On the other hand, under the nonlinear effect limit condition, an attainable transmission distance, $L_{nonlinear}$, decreases as the amplifier spacing increases at a specific amplifier output power, and decreases as the amplifier output power increases at a specific amplifier spacing below around 40 km, as shown in Fig. 6.6. Therefore, if the amplifier output power is given, L_{noise} and $L_{nonlinear}$ coincide with each other at a characteristic amplifier spacing.

In Fig. 6.7, solid lines show the coinciding transmission distance at 2.5 and 10 Gbit/s, which is regarded as an expected transmission distance considering both the limits, and oblique broken lines indicate the characteristic amplifier spacing.

In the model of cascaded amplifier systems shown in Fig. 6.1, loss of optical components included in an optical amplifier is not considered. If the optical

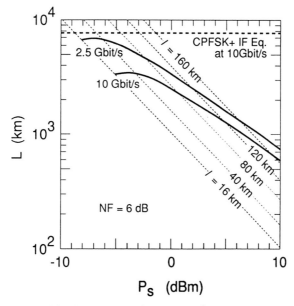

Fig. 6.7 Expected system performance.

signal transmittance, η_a, in optical components, for example isolators, filters and couplers, equipped at amplifier input and/or output ports is considered, $(e^{\alpha l} - 1)$ should be replaced by $(e^{\alpha l} - \eta_a)$ in equations 6.1, 6.2 and 6.4. This has an effective influence upon system performance only when the amplifier gain $G\, (= e^{\alpha l})$ is small. When $\eta_a = 1$, the attainable transmission distance is a maximum at the amplifier spacing of 0 km, while the spacing l_0 to achieve a maximum transmission distance becomes finite, when $\eta_a \neq 1$. For example, assuming $\eta_a = 0.5$ (3 dB loss), $l_0 = 16$ km.

Applying CPFSK modulation and heterodyne delay detection to cascaded amplifier systems, both the noise accumulation limit and the nonlinear effect limit are dominant compared with the fiber dispersion limit below a few tens of Gbit/s. A maximum transmission distance is obtained at an amplifier spacing of around 40 km. An optimum amplifier output power is as small as -10 to 0 dBm. A few thousands to ten thousands of kilometers are the expected transmission distances in a Gbit/s range.

6.1.7 Measured error rate performance

The performance estimation mentioned above can be verified by the results of a long distance transmission experiment using CPFSK modulation and heterodyne delay detection at 2.5 Gbit/s (Saito, et al., 1992).

In the experiment, twenty-four erbium-doped fiber amplifiers (EDFAs) with a spacing of 100 km were employed. The total transmission distance was 2500 km. The transmission fibers were dispersion-shifted single-mode fibers with a zero dispersion wavelength of around 1.5505 μm. The average dispersion slope was 0.063 ps/km/nm². The average loss was 0.21 dB/km at the optical signal wavelength of 1.552 μm. The mode field diameter for the fibers was 8 μm. The EDFAs used were backward pumped by 1.48 μm semiconductor lasers. The EDF was co-doped with aluminum. The erbium and aluminum concentrations were 1000 ppm and 5000 ppm, respectively, and both were concentrated in the center of the core. The length of EDF was adjusted to 48 m in order to obtained an output power of higher than 10 dBm and a low noise figure. Configurations and performance of the sender and receiver used were similar to those described in Chapter 7.

Measured error rate performance for four different transmission systems of 1100 km-long, 1600 km-long, 2100 km-long and 2500 km-long is shown in Fig. 6.8. EDFA output and input power were around 2 dBm and −19 dBm, respectively, and the noise figure was 8.2 dB on average. As a result, P_s/NF was about −6 dBm. In Fig. 6.8, an error rate floor appears in the curves except the back-to-back curve. No error rate floor is found in the back-to-back error rate performance, even though the slope degrades slightly from the shot noise limit curve, which is calculated by assuming a detector quantum efficiency η of 1. One origin of the error rate floor is the nonlinear effects.

Power penalties, which are defined as the optical power increase required to compensate the transmission performance degradation and to obtain the error

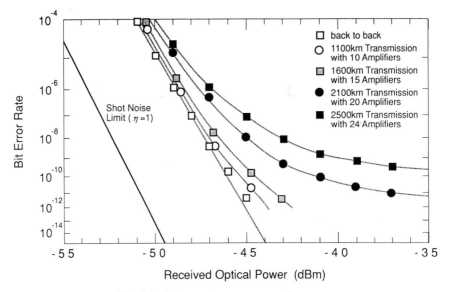

Fig. 6.8 Measured error rate performance.

IN-LINE AMPLIFIER SYSTEMS

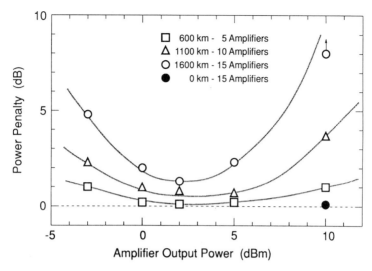

Fig. 6.9 Power penalty as a function of amplifier output power.

rate of 10^{-9}, are shown in Fig. 6.9 as a function of the amplifier output power. An open circle with an arrow shows that the power penalty is higher than the level indicated. A solid circle shows the power penalty in the case when all the fibers were replaced by optical attenuators with the same loss as the fibers. As both the amplifier output power and the transmission fiber length increase, power penalties increase. Therefore, a power penalty increase found in the right part of

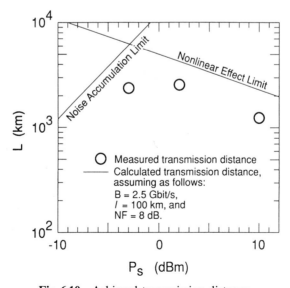

Fig. 6.10 Achieved transmission distance.

the figure originates from the nonlinear effect, while that found in the left part is caused by the noise accumulation.

Transmission distances achieved by using amplifiers with three different levels of the output power are shown by open circles in Fig. 6.10. Those points plotted the datum representing an error rate of 10^{-9}. The gain of the amplifiers used was 21 dB, which corresponded to the amplifier spacing of 100 km. Their noise figure was 8 dB. Transmission distances calculated by using those experimental parameters are also shown in the figure. There was an optimum amplifier output power to achieve the maximum distance, and either the lower or the higher output power gave a shorter transmission distance. Shortening of the distance in a lower power region is caused by the noise accumulation, and that in a higher power region resulted from the nonlinear effect. There was some discrepancy between the observed datum and the calculation

6.1.8 Factors affecting transmission performance

In order to operate and maintain a cascaded amplifier system in the field, a number of functions including remote control, supervision, fault location, and loop back are required. To facilitate these functions, some optical components were inserted into the optical signal path of the optical amplifiers. This degrades the amplifier performance, that is the overall P_s/NF. The degradation, however, will not fatally affect system performance. It will simply increase the degradation allocation in system design.

There are many kinds of imperfections in optical devices and components forming an optical amplifier. For example, the need for a wide optical filter in order to reject the amplified spontaneous emission accumulation; the broad spectral width of the laser diodes; and the reflection originating at splicing or connecting points. These imperfections increase the noise and decrease the maximum attainable amplifier spacing. The influence of the wide width has already been quantified, and it will also result in increased degradation allocation in system design. The influence of the broad spectral width should be further studied from the viewpoint of both spectral broadening in each casaded amplifier, and the carrier phase fluctuations caused by the multiplier effect of carrier fluctuations and fiber dispersion.

Aging of optical devices such as pumping laser diodes and dielectric multi-layer thin film optical filters may affect system performance. The stable operation of a pumping laser diode for over 5000 hours in a high temperature atmosphere was confirmed both at 1.48 µm and 0.98 µm wavelengths. On the other hand, a lot of experience has been gained with thin film optical filters used as polarization beam couplers in many 1.3 µm submarine transmission systems installed by KDD or NTT. Such filters have been operating without a problem for over four years in the longest case. A problem is the adhesive used in the optical path. Even though most adhesive films have very little absorption loss (0.01 dB or less), heat generated by the absorption of high optical power degrades the film and causes an abrupt loss increase. Therefore, use of adhesive in the optical path should be avoided wherever possible.

6.2 COMMON AMPLIFIERS

Kyo Inoue

Since optical amplifiers can simultaneously amplify multiplexed optical signals, they are very useful in optical FDM systems. In this section, optical amplifiers are described from the viewpoint of FDM transmission applications. The important characteristics of optical amplifiers for FDM systems are the signal gain spectrum, mutual gain saturation, and interchannel crosstalk. We will discuss these characteristics for LD amplifiers and fiber amplifiers. Historically, LD amplifiers were the first to be studied while fiber amplifiers are a recent development.

6.2.1 LD amplifier

(a) Gain spectrum

Typical signal gain spectra of a bulk-type LD amplifier are shown in Fig. 6.11(a) given by Mukai, Inoue and Saitoh (1987a). Shown in Fig. 6.11(b) are the amplified spontaneous emission (ASE) spectra, which are equivalent to the gain spectrum of an MQW-type LD amplifier as in Magari, *et al.* (1990). The 3 dB gain bandwidth is 50 nm for the bulk amplifier and 100 nm for the MQW amplifier. Input signals within the bandwidth are simultaneously amplified.

These large bandwidths are attractive for FDM transmission systems. Another advantage of LD amplifiers is that arbitrary gain peak wavelengths can be obtained by choosing an appropriate semiconductor material. The larger bandwidth of the MQW-type amplifier is due to the unique band structure of the MQW medium.

(b) Mutual gain saturation

When the total power of multiplexed light inputs to an optical amplifier is small, each light is amplified with an unsaturated gain value. For large input powers, on the other hand, the gain of one channel is affected by the other channels, i.e. mutual gain saturation occurs.

Mutual gain saturation in an LD amplifier was measured as shown in Fig. 6.12, from Mukai, Inoue and Saitoh (1987b). Here, signal gain as a function of amplifier

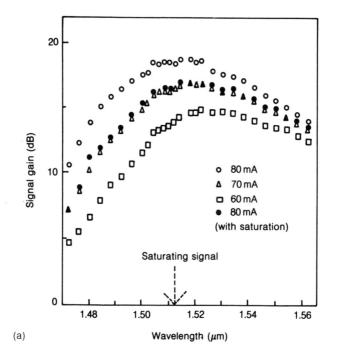

(a)

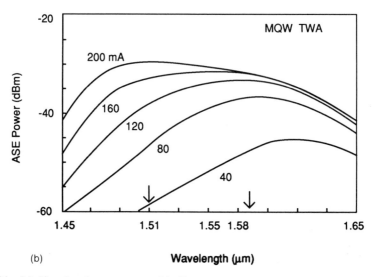

(b)

Fig. 6.11 (a) Signal gain spectrum of bulk-type LD amplifier*. (b) ASE spectrum of MQW-type LD amplifier[†].
*Source: Mukai, *et al.*, 1987a.
[†]Source: Magari, *et al.*, 1990.

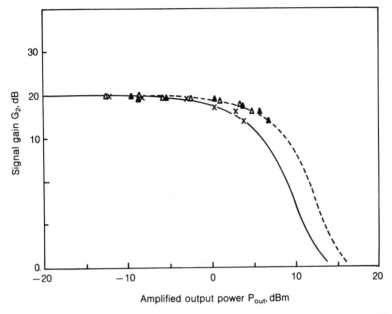

Fig. 6.12 Signal gain saturation for two-channel amplification in LD amplifier. Source: Mukai, *et al.*, 1987b.

output power is plotted for two-channel amplification with an equal input power of each channel.

Signal gain begins to reduce at lower output power when two lights are simultaneously amplified, marked ($\times$), compared to one channel amplification ($\triangle$). This figure also shows that signal gain as a function of total output power ($\blacktriangle$) fits the saturation curve for one-channel amplification. This means that the gain saturation characteristic is determined by the total power and signal gain in multichannel amplification is readily obtained from the gain saturation curve for one-channel amplification. When the channel separation is narrow enough such that saturation dependence on wavelength can be ignored, the one-channel saturation output power in N-channel amplification becomes 1/N of that in one-channel amplification.

(c) Interchannel crosstalk

The mutual gain saturation described above directly induces interchannel crosstalk in ASK transmission where the total light power randomly changes due to the signal modulation of each channel. Under the gain saturated condition, the signal gain changes with the total light power, resulting in interchannel crosstalk.

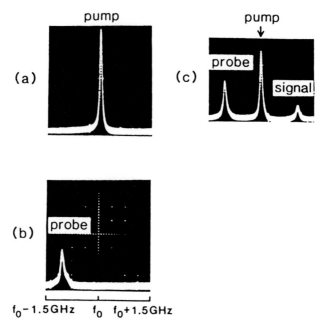

Fig. 6.13 Measurement of four-wave mixing in LD amplifier: (a) pump light is injected; (b) probe light is injected; (c) pump and probe lights are injected. Additional light, signal, is observed in (c).
Source: Inoue, *et al.*, 1987.

Fortunately, this type of crosstalk does not occur in power-constant modulation schemes, such as FSK and PSK.

Another mechanism that induces crosstalk in LD amplifiers is four-wave mixing or carrier density modulation described in Inoue, Mukai and Saitoh (1987) and Darcie, Jopson and Tkach (1987). An experimental result for the four-wave mixing in an LD amplifier is shown in Fig. 6.13 from Inoue, Mukai and Saitoh (1987), where output spectra of an LD amplifier are observed. When two lights of different frequencies, pump and probe, are input to an LD amplifier, additional light of a different frequency, signal, appears at the amplifier output as shown in Fig. 6.13(c). This phenomenon can be understood as follows.

When two lights of different frequencies are input to an LD amplifier, total light intensity within the amplifier vibrates at the beat frequency of the two lights. Under the gain saturated condition, the electron carrier density in the LD amplifier changes with the total light intensity. Thus, the electron carrier density vibrates at the beat frequency. When the electron carrier density changes, both the refractive index and the medium gain change as well. Therefore, the input lights are phase- and amplitude-modulated at the beat frequency. Thus, a modulation side-band wave is observed as the additional light.

The additionally generated light induces interchannel crosstalk and degrades

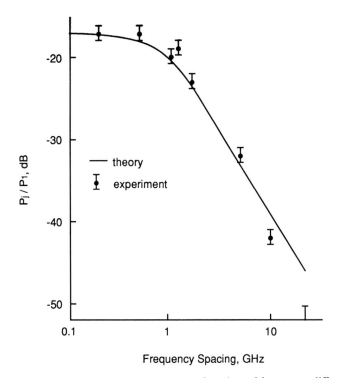

Fig. 6.14 Four-wave mixing efficiency as a function of frequency difference. Source: Grosskopf, et al., 1988.

system performance even in power-constant modulation schemes. Fortunately, the four-wave mixing efficiency rapidly decreases as the frequency separation of input lights increases, as shown in Fig. 6.14, where the power ratio of an original light (P1) and a four-wave mixing light (P3) is plotted as a function of frequency separation as in Grosskopf, et al. (1988). This is because the electron carrier density cannot follow high beat frequencies. Thus, four-wave mixing is avoided when a frequency separation of more than several GHz is chosen in FDM systems.

(d) System experiments

Several transmission experiments have been reported on common amplification with an LD amplifier. One experiment saw the amplification of 10-channel FSK signals with a channel spacing of 8 GHz (Shibutani, et al., 1989). The channel separation was large enough so that the error rate degradation due to four-wave mixing was not induced.

The influence of four-wave mixing was observed in other experiments that used narrower channel separation. Receiver sensitivity as a function of amplifier input

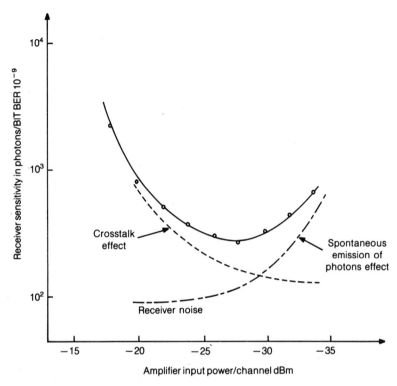

Fig. 6.15 Receiver sensitivity degradation due to four-wave mixing in LD amplifier. Source: Glance, *et al.*, 1989.

power was measured for a 200 MHz-spaced 4-channel FSK system. The result is shown in Fig. 6.15 from Glance, *et al.* (1989). It shows that sensitivity is degraded by amplifier noise at small input powers; four-wave mixing degrades sensitivity at large input powers. A power penalty caused by four-wave mixing was also observed in 3-channel amplification of 2 GHz-spaced FSK signals as in Ryu, Mochizuki and Wakabayashi (1989).

6.2.2 Fiber amplifiers

(a) Gain spectrum

In the first study of fiber amplifiers, erbium Er^{3+} was simply doped into conventional silica-based fibers. This type of fiber amplifier has a narrow gain bandwidth (several nm) and is not suitable for FDM systems. Afterwards, it was found that the gain spectrum broadened when Al was codoped into the fiber together with Er. Appropriate host materials were used as described in Atkins,

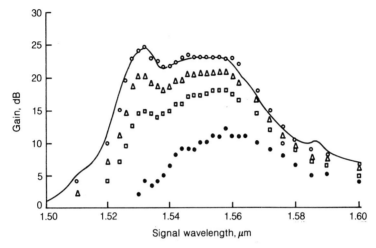

Fig. 6.16 Signal gain spectra of fiber amplifier.
Source: Atkins, *et al.*, 1989.

et al. (1989) and Fevrier, *et al.* (1989). An example of such a gain spectrum is shown in Fig. 6.16, where the amplifier was biased at various pumping conditions described in Atkins, *et al.* (1989). The host material is SiO_2–Al_2O_3–P_2O_5. A 3 dB bandwidth of 35 nm was obtained, which is the largest value for a fiber amplifier to date.

It is shown in Fig. 6.16 that the gain spectrum of fiber amplifiers is dependent on the pumping condition. Another measurement showed that it also depends on fiber length and input power. Thus, it should be noted that the gain spectrum of fiber amplifiers is not uniquely determined, depending on parameters such as pumping power, fiber length, and signal input power.

(b) Mutual gain saturation

In the gain saturated condition, the signal gain of one channel is affected by the other channels (Desurvire, Giles and Simpson, 1989). The measured mutual gain saturation characteristic for two-channel amplification is shown in Fig. 6.17, where the wavelength difference was 0.5 nm (Inoue, *et al.*, 1991). As in LD amplifiers, signal gain begins to saturate at a lower output power in two-channel amplification, shown as △, compared to the one-channel case, (○), and the signal gain curve as a function of total output power, (×), is matched to that in one-channel amplification. Thus, the saturation output power for one channel becomes 1/N in a narrow spaced N-channel amplification.

The wavelength range of this feature depends on fiber material. Al/Er-doped fiber amplifiers have a wavelength range of at least several nm according to Inoue,

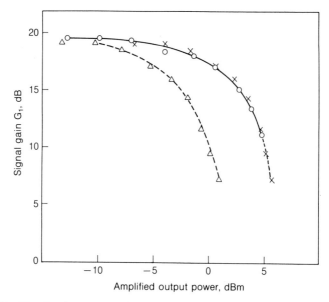

Fig. 6.17 Signal gain structure for two-channel simplification in fiber amplifier. Source: Inoue, et al., 1989.

Toba and Nosu (1991). For a large channel separation, however, the saturation characteristic is not so simply defined due to the dependence of saturation on wavelength and the gain broadening characteristic.

(c) Interchannel crosstalk

Mutual gain saturation can induce interchannel crosstalk in ASK systems. The crosstalk characteristic of a fiber amplifier was investigated for two-channel amplification as shown in Fig. 6.18, where one of the two channels was intensity-modulated in Pettitt, Hadjifotiou and Baker (1989). The amplifier gain was set at 20 dB (×), 19 dB (+), and 15 dB (○). Though crosstalk occurs at a low modulation frequency, it decreases as modulation frequency increases. For a modulation frequency of larger than 10 kHz, the crosstalk level is negligible. Thus, crosstalk is not a problem in practical systems if the modulation rate is high enough.

The above crosstalk characteristic is attributed to the response time of the fiber amplifier. When the input light intensity changes much more rapidly than the response time, the amplifier responds to only the averaged input power and the signal gain becomes constant. Thus, no interchannel crosstalk occurs. The response time of a fiber amplifier was measured under two-channel amplification where one channel was emitted from a strong saturating laser and the other was from a weak probe laser (Giles, Desurvire and Simpson, 1989). The result is shown

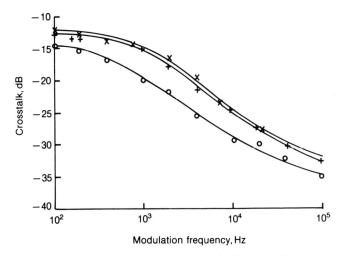

Fig. 6.18 Crosstalk measurement in fiber amplifier.
Source: Pettitt, *et al.*, 1989.

in Fig. 6.19, where the signal outputs are measured for a pulse input of the saturating laser. The response time was observed to be several hundreds of μs. Channel crosstalk is not incurred with a modulation frequency faster than this response time, which matches the results shown in Fig. 6.18.

In LD amplifiers, four-wave mixing is another crosstalk mechanism. It results from a beat vibration of the electron carrier density. In fiber amplifiers, on the other hand, the gain medium cannot respond to a beat vibration of several GHz as described above. Thus, this type of crosstalk, as well as crosstalk due to mutual gain saturation, does not occur in fiber amplifiers.

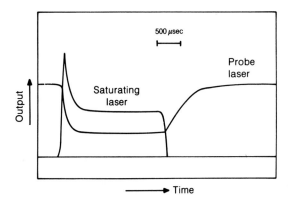

Fig. 6.19 Dynamic characteristic of fiber amplifier.
Source: Giles, *et al.*, 1989.

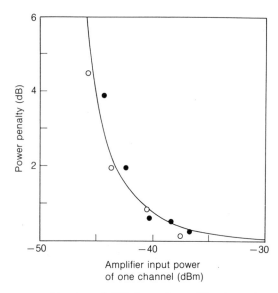

Fig. 6.20 Power penalties 1- and 100-channel amplification using fiber amplifier. Source: Inoue, et al., 1991.

(d) System experiments

Several system experiments have been reported using an Er^{3+}-doped fiber amplifier. Welter, et al. (1989) carried out 16-channel common amplification in a 10 GHz-spaced FSK heterodyne system. Inoue, et al. (1991) reported 100-channel amplification in a 10 GHz-spaced FSK direct detection system. The latter experiment found that the characteristic of error-rate degradation due to fiber amplifiers is the same for one- or 100-channel amplification as shown in Fig. 6.20, where open and solid circles are power penalties in 100-channel and one-channel amplification experiments. Thus, it was confirmed that amplifier performance for multichannel amplification can be estimated from that measured with only one channel.

6.2.3 System design

The above subsections describe the basic characteristics of optical amplifiers used for multichannel systems. In this subsection, designing multichannel systems using optical amplifiers is described. In particular, we discuss the channel capacity. There are three factors limiting the channel capacity: signal gain bandwidth, crosstalk and amplifier noise.

The allowable number of channels limited by gain bandwidth is readily obtained from the signal gain spectrum; roughly speaking: (3 dB gain bandwidth)/(channel spacing) = (channel capacity) from the viewpoint of gain bandwidth. It should be noted, however, that signal level difference is accumulated in FDM systems with multi-stage amplifiers. Available wavelength bandwidth is narrow in the total system, compared to that of one amplifier. In order to overcome this, several techniques to equalize signal levels are proposed in Tachibana, *et al.* (1991) and Inoue, Kominato and Toba (1991).

There are two mechanisms of channel crosstalk in LD amplifiers: mutual gain saturation and four-wave mixing. The former crosstalk is avoided in power-constant modulation systems, and the latter crosstalk can be avoided if the channel spacing exceeds more than several GHz. For fiber amplifiers, on the other hand, there is no such limit.

The limitation due to amplifier noise is considered as follows. When the amplifier input power is small, the output signal ratio to amplifier noise power (ASE power) is small; thus receiver sensitivity is lower than if no amplifier is used. Provided that an allowable penalty is specified, the allowable minimum input power of one channel can be estimated from noise calculations for one-channel amplification as in Olsson (1989). On the other hand, total power input to the amplifier is limited due to the signal gain saturation. For large input power, signal gain is small. The maximum total power is determined from the signal gain required by the system design. Once the maximum total input power is obtained, channel capacity is estimated as: (the allowable maximum power of total input)/(the allowable minimum power per channel) = (allowable number of

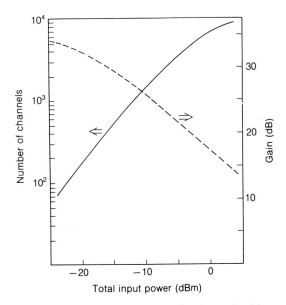

Fig. 6.21 Calculated example of channel capacity determined by amplifier noise. Source: Inoue, *et al.*, 1991.

channels). A calculated example for an FSK direct detection system using a fiber amplifier is shown in Fig. 6.21.

In designing an FDM system with optical amplifiers, the above factors must be considered as a whole.

REFERENCES

Atkins, C. G., Massicott, J. F., Armitage, J. R., Wyatt, R., Ainslie, B. J. and Craig-Ryan, S. P. (1989) *Electronics Letters*, **25**/14, p. 910.

Bergano, N., Aspell, J., Davidson, C., Trischitta, P., Nyman, B. and Kerfoot, F. (1991) *Technical Digest of Optical Fiber Communication Conference 1991*, **PD13**.

Darcie, T. E., Jopson, R. M. and Tkach, R. W. (1987) *Electronics Letters*, **23**/25, p. 1392.

Desurvire, E., Giles, C. R. and Simpson, J. R. (1989) *IEEE Journal Lightwave Technology*, **7**/12, p. 2095.

Elrefaie, A. F., Wagner, R. E., Atlas, D. A. and Daut, D. G. (1988) *IEEE/OSA Journal of Lightwave Technology*, **6**/5, pp. 704–709.

Fevrier, H., Auge, J., Parlier, V., Bousselet, P., Dursin, A., Marcerou, J. F. and Jacquier, B. (1989) *Proceedings ECOC'89*.

Giles, C. R., Desurvire, E. and Simpson, J. R. (1989) *Optics Letters*, **14**/16, p. 880.

Glance, B. S., Eisenstein, G., Fitzgerald, P. J., Pollock, K. J. and Raybon, G. (1989) *IEEE/OSA Journal Lightwave Technology*, **7**/5, p. 759.

Gordon, J. and Haus, H. (1986) *Optics Letters*, **11**/10, pp. 665–667.

Gordon, J. and Mollenauer, L. (1990) *Optics Letters*, **15**/23, pp. 1351–1353.

Grosskopf, G., Ludwig, R., Waarts, R. G. and Weber, H. G. (1988) *Electronics Letters*, **24**/1, p. 31.

Inoue, K., Kominato, T. and Toba, H. (1991) *IEEE Photonics Technology Letters*, **3**/8, p. 718.

Inoue, K., Mukai, T. and Saitoh, T. (1987) *Applied Physics Letters*, **51**/14, p. 1051.

Inoue, K., Toba, H. and Nosu, K. (1991) *IEEE/ISO Journal Lightwave Technology*, **9**/3, p. 368.

Inoue, K., Toba, H., Shibata, N., Iwatsuki, K., Takada, A. and Shimizu, M. (1989) *Electronics Letters*, **25**/9, p. 594.

Ito, T. (1991) *Technical Digest of Topical Meeting on Optical Amplifiers and Their Applications*, **ThA1**, pp. 78–81.

Iwashita, K. and Takachio, N. (1990) *IEEE/OSA Journal of Lightwave Technology*, **8**/3, pp. 367–375.

Magari, K., Kondo, S., Yasaka, H., Noguchi, Y., Kataoka, T. and Mikami, O. (1990) *IEEE Photonics Technology Letters*, **2**/11, p. 792.

Mollenauer, L. F., Lichtman, E., Harvey, G. T., Neubelt, M. J. and Nyman, B. M. (1992) *Technical Digest of Optical Fiber Communication Conference 1992*, **PD10**, pp. 351–354.

Mollenauer, L., Nyman, B., Neubelt, M., Raybon, G. and Evangelides, S. (1991) *Electronics Letters*, **27**/2, pp. 178–179.

Mukai, T., Inoue, K. and Saitoh, T. (1987a) *Applied Physics Letters*, **51**/6, p. 381.

Mukai, T., Inoue, K. and Saitoh, T. (1987b) *Electronics Letters*, **23**/8, p. 396.

Olsson, N. A. (1989) *IEEE/OSA Journal of Lightwave Technology*, **7**/7, p. 1071.

REFERENCES

Pettitt, M. J., Hadjifotiou, A. and Baker, R. A. (1989) *Electronics Letters*, **25/6**, p. 416.

Ryu, S., Mochizuki, K. and Wakabayashi, H. (1989) *IEEE/OSA Journal Lightwave Technology*, **7/10**, p. 1525.

Saito, S., Aiki, M. and Ito, T. (1993) *IEEE/OSA Journal of Lightwave Technology*, **11/2**, pp. 331–342.

Saito, S., Imai, T. and Ito, T. (1991) *IEEE/OSA Journal of Lightwave Technology*, **9/2**, pp. 161–169.

Saito, S., Murakami, M., Naka, A., Fukada, Y., Imai, T., Aiki, M. and Ito, T. (1991) *Post-deadline papers of 17th European Conference on Optical Communication/8th International Conference on Integrated Optics and Optical Fiber Communications*, **A.PDP.5**, pp. 68–71.

Saito, S., Murakami, M., Naka, A., Fukada, Y., Imai, T., Aiki, M. and Ito, T. (1992) *IEEE/OSA Journal of Lightwave Technology*, **10/8**, pp. 1117–1126.

Shibutani, M., Cha, I., Yamazaki, S., Kitamura, M. and Emura, K. (1989) *Proceedings IOOC'89*, **21B4–5**.

Tachibana, M., Laming, R. I., Morkel, P. R. and Payne, D. N. (1991) *IEEE Photonics Technology Letters*, **3/2**, p. 118.

Taga, H., Edgawa, N., Yoshida, Y., Yamamoto, S., Suzuki, M. and Wakabayashi, H. (1992) *Technical Digest of Optical Fiber Communication Conference 1992*, **PD12**, pp. 359–362.

Welter, R., Laming, R. I., Sessa, W. B., Vodhanel, R. S., Maeda, M. W. and Wagner, R. E. (1989) *Electronics Letters*, **25/20**, p. 1333.

7
System applications

7.1 LONG-HAUL TRUNKS AND SUBMARINE COMMUNICATIONS

Takeshi Ito

Multi-Gbit/s long-span transmission systems, focusing on a repeaterless coherent transmission system, will be described. First, the limits of fiber input power limit and receiver sensitivity limit are discussed. These are the key factors for repeaterless systems. Second, the merits of coherent transmission in repeaterless systems from the viewpoint of transmission performance are summarized. Next, engineered transmission equipment designed by NTT laboratories and a field trial carried out with the equipment are introduced. Finally, the variations in the characteristics of coherent laser diodes caused by aging will be dicussed.

7.1.1 Technologies for long-span transmission systems

High sensitivity receiving and high output power dominate the performance of repeaterless systems. Receiver sensitivity is limited by shot noise. Two technologies are promising to approach the shot noise limit. One is the utilization of erbium-doped fiber amplifiers as preamplifiers, and the other is coherent detection.

The state of the art in high sensitivity reception is shown in Table 7.1 for the range 0.6 to 10 Gbit/s. Sensitivity is expressed as photons/bit. The upper column is for coherent detection and the lower column is for optical preamplifiers. The record for coherent detection was achieved by AT&T (Kahn, 1989 and Kahn, *et al.*, 1990) and NTT (Imai, *et al.*, 1990 and Iwashita and Norimatsu, 1991). These four records for optical preamplifiers were achieved by Alcatel (Gabla, *et al.*, 1992) and NEC (Saito, *et al.*, 1991 and 1992). Sensitivity with the optical preamplifier is defined as the input power to an erbium-doped fiber amplifier, but not by the input power to an erbium-doped fiber. The difference between the two corresponds to the front end loss. All values were achieved with the 0.98 µm wavelength forward or backward pumping configuration. From these records, at present, the sensitivity

Table 7.1 Reported sensitivity record

Transmission bit-rate (Gbit/s)	0.622	1	2.5	4	5	10
Coherent						
Sensitivity (photons/bit)	—	46	67	72	—	121
Mod./demod. scheme		PSK/ Homodyne	CPFSK/ Heterodyne	PSK/ Homodyne		PSK/ Homodyne
Pre-EDFA						
Sensitivity (photons/bit)	136	—	115	—	151	268

of coherent detection is 2.5 to 3.5 dB higher than that obtained from optical preamplifiers in the multi-Gbit/s region.

Erbium-doped fiber amplifiers are very effective to obtain high output optical power. However, it should be noted that the optical power that can be utilized in a repeaterless system is limited by the fiber input power limit, but not by the optical amplifier output. The maximum usable power in repeaterless systems is limited by the stimulated Brillouin scattering (SBS), which is a kind of nonlinear effect, as discussed in detail in section 3.4.

With SBS, the backscattered optical power at the fiber entrance end abruptly increases with the optical input power, and the transmitted optical power at the fiber output end does not increase in proportion to the optical input power. It is usually caused by a highly coherent and strong lightwave. It is usually observed at optical fiber input powers higher than 7 dBm, if the input light is a continuous wave, its spectral width is less than several MHz, and the fiber length is longer than 30 km. However, the threshold in the stimulated Brillouin scattering increases when the fiber input light is modulated. This is because the input light becomes less coherent due to modulation, and it causes less of a nonlinear effect.

If the input light is modulated by random signals, the threshold is higher than 20 dBm. The observed spectra of the backscattered light are shown in Fig. 7.1, when the light is modulated by continuous phase frequency shift keying at 2.5 Gbit/s. This figure is obtained by modulating the light with a modulation index of 1.0. The spectrum shows two remarkable peaks. The optical input power was 18.5 dBm. The stimulated Brillouin scattering exceeds the Rayleigh scattering. On other hand, in the case of a light modulated with a modulation index of 0.7, the stimulated Brillouin scattering is negligible compared to the Rayleigh scattering. This is because the spectrum is much broader. Note that there is considerable overlap in the Rayleigh scattered light and the stimulated Brillouin scattered light.

The optical fiber input limit depends on modulation conditions such as a transmission bit-rate, modulation scheme, and so on. The fiber input power limit is defined by the threshold given by stimulated Brillouin scattering. The threshold

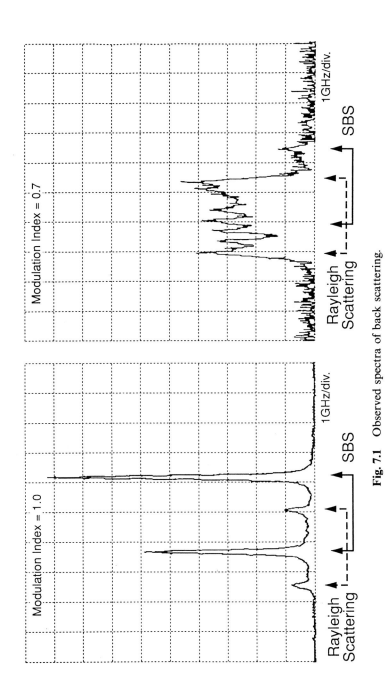

Fig. 7.1 Observed spectra of back scattering.

for amplitude shift keying modulation, similar to intensity modulation, does not depend on transmission bit-rate. This is because the carrier lightwave is not suppressed, even though it is modulated, and the carrier lightwave is highly coherent. On the other hand, the threshold for coherent modulation such as frequency shift keying or phase shift keying increases with the transmission bit-rate.

The stimulated Brillouin scattering effect is most suppressed at a modulation index of around 0.65 with frequency shift keying according to Sugie (1991). The maximum suppression is also obtained with phase shift keying modulation. This is because these modulation schemes yield sufficient spectral broadening.

7.1.2 Transmission distance of repeaterless system

Considering both the receiver sensitivity limit and the fiber input power limit as well as the fiber dispersion limit, expected transmission distance for repeaterless systems is shown in Fig. 7.2. Compared with IM/DD systems, coherent systems have three major merits, as follows. The first is that coherent detection has high sensitivity, although the advantage is not great. Therefore, the transmission distance can be extended by at least 20 km. The second is that the attainable transmission distance does not decrease as the transmission bit-rate increases. This is because the fiber input power limit increases as the bit-rate increases, as just discussed. The third is that heterodyne detection can compensate the fiber dispersion by means of intermediate frequency delay equalization, as shown in

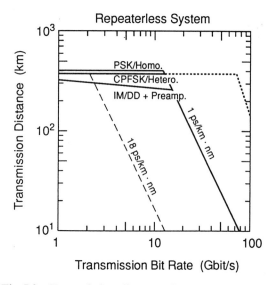

Fig. 7.2 Transmission distance of repeaterless system.

LONG-HAUL TRUNKS AND SUBMARINE COMMUNICATIONS 221

detail in section 3.5. Therefore, heterodyne detection systems are applicable both to 1.3 μm zero-dispersion wavelength fiber systems and 1.55 μm zero-dispersion wavelength fiber systems in the multi-Gbit/s region. In addition, they can be applied to 10 Gbit/s or higher repeaterless transmission.

If an optical amplifier is utilized as a booster amplifier, a system gain of higher than 60 dB is realized at 2.5 Gbit/s. And it is applicable both to 1.3 μm zero-dispersion wavelength fiber and to 1.55 μm zero-dispersion wavelength fiber.

7.1.3 Outline of experimental repeaterless transmission system

In order to confirm the feasibility of a repeaterless coherent transmission system, engineered transmission equipment was designed as recorded by Hayashi (1991). The outline of the target repeaterless coherent transmission system is shown in Table 7.2. The transmission bit-rate was 2.5 Gbit/s corresponding to synchronous transport module level 16, that is STM-16. The wavelength was 1.545 μm. The modulation scheme was continuous phase frequency shift keying with a modulation index of 0.8. This was decided from the viewpoints of signal bandwidth reduction and sensitivity degradation minimization. Optical signals were converted into intermediate frequency signals of 5 GHz by heterodyne detection. The intermediate frequency signals are demodulated to baseband signals by delay detection. The output power target was higher than 9 dBm, and receiver sensitivity for an error rate of 10^{-11} was less than -41 dBm. Therefore, the system gain was expected to be more than 50 dB.

The configuration of the prototype system is schematically shown in Fig. 7.3. Baseband signals were fed to the input of the transmitter, and converted into optical signals by a laser diode. The laser diode was a long cavity three-electrode DFB-LD which had flat response characteristics in frequency modulation and a spectral width of less than 2 MHz. After being transmitted through a long fiber cable, optical signals were fed to the receiver input. Polarization of the optical signals fluctuated along the fiber.

In order to compensate for these fluctuations, a polarization diversity circuit was applied. That is, input optical signals were divided into two orthogonal linear

Table 7.2 Outline of prototype repeaterless transmission system

Transmission bit-rate	2.48832 Gbit/s
Wavelength	1.545 μm
Modulation	CPFSK (modulation index = 0.8)
Detection	Heterodyne/delay detection
Intermediate frequency	4.98 GHz
Output power	$> +9$ dBm
Receiver sensitivity	< -41 dBm ($P_e = 10^{-11}$)
System gain	> 50 dB

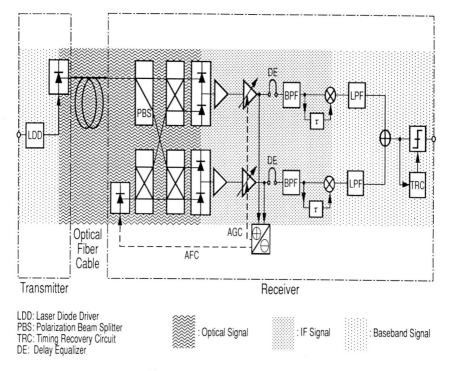

Fig. 7.3 System configuration.

polarization components by a polarization beam splitter. The two signals were mixed with local optical power under the polarization matched state by a balanced dual PIN photodiode, and converted to intermediate frequency signals. Each signal was amplified under automatic gain control, and shaped by a 5th Thomson band-pass filter. They were then demodulated to baseband signals, which were summed at an adder circuit and regenerated. The delay equalizers were adaptively used if necessary. For example, when optical signals were transmitted through a 1.3 µm zero-dispersion wavelength fiber longer than 200 km, the equalizers were employed.

In the heterodyne detection system, a local laser diode frequency had to be slaved to that of the transmitting laser diode by an automatic frequency control. The frequency difference between the two laser diodes, that is the intermediate frequency, was kept constant. However, there were two intermediate frequency components. Their amplitudes fluctuated with the polarization fluctuations. At any one instant, one of them was zero. Therefore, the two components were compared, and the stronger one was used as a reference by the automatic frequency control circuit. As a result, the frequency variation was stabilized below 1 MHz. On the other hand, the gain of the intermediate frequency amplifiers was automatically controlled by the sum of the two components. Therefore, an optical

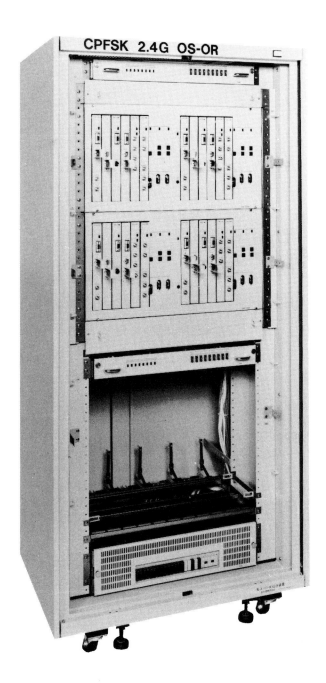

Fig. 7.4 Overall view of transmission equipment.

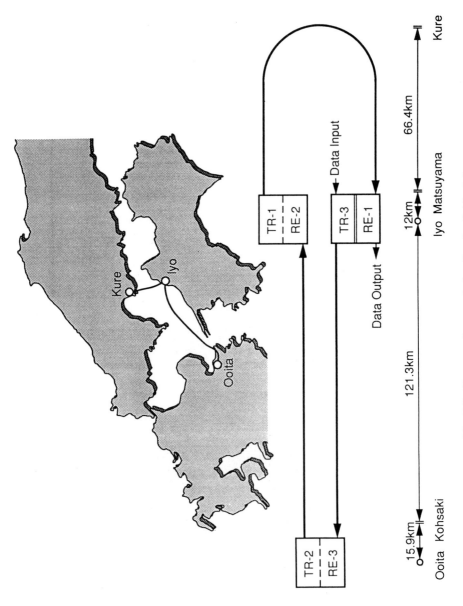

Fig. 7.5 Route and transmission line for field trial.

signal ranging from −45 dBm to −30 dBm was stabilized with a gain variation within ±1 dB. The local laser diode used here was also a long cavity three-electrode DFB-LD. Its wavelength was electronically tunable to within ±0.4 nm and its spectral width was less than 1 MHz.

An overall view of the prototype repeaterless transmission equipment is shown in Fig. 7.4. It was 27 cm high, 33 cm wide and 34 cm deep. Total consumption power was 350 W. The equipment was cooled by forced air. The transmitter was composed with one panel, while the receiver was done with five panels. The optical signals were fed to the optical connector in front of the first receiver panel, that is, the front-end circuit panel, and converted to the two intermediate signals at the panel. Each signal was then transmitted through the coaxial lines from left to right. The two signals were demodulated to the baseband signals and summed at the fourth receiver panel, that is, the demodulation circuit panel. The baseband signals were regenerated at the fifth receiver panel, that is, the regenerator circuit panel. The second receiver panel was the intermediate frequency amplifier. An automatic frequency control signal was returned through this coaxial line from the intermediate frequency amplifier panel to the front-end circuit panel. The third receiver panel was the intermediate frequency delay equalizer and used adaptively.

A field trial was carried out using three prototype equipment sets. The trial used a round route connecting the three stations of Kure, Iyo and Ooita, as shown in Fig. 7.5. These are towns bordering the Inland Sea of Japan, Existing commercial cable was used, and two of the six cores were used in the trial. The fiber length totaled 431 km. More than 85% of it was submarine cable of 1.55 μm zero-dispersion wavelength fiber, and the rest was land cable of 1.3 μm zero-dispersion wavelength fiber. Total loss for the section TR-1/RE-1 was 44.0 dB. Losses for the sections TR-2/RE-2 and TR-3/RE-3 were 33.7 and 33.1 dB.

7.1.4 Measured polarization fluctuation and error rate variation

Some of the results obtained in the field trial are described. The relationship between polarization state fluctuation and bit error rate variation over 30 minutes is shown in Fig. 7.6. Received optical power was −43 dBm. The measured polarization state fluctuated with polarization power ratios ranging from 0.0 to 1.0. Polarization power ratio was defined as the ratio of parallel polarization power to total optical power. The observed fluctuation was faster than usual because the end of the fiber connected to the receiver was jarred by the movement of operational personnel. The measured bit error rate variation corresponded to the received optical power variation of less than 0.5 dB, and was almost the same as the bit error rate variation observed in a polarization fluctuation free state. These facts proved the effectiveness of polarization diversity techniques.

After preventing the operational personnel from jarring the fiber, polarization state fluctuations were measured over 106 hours and are shown in Fig. 7.7. The

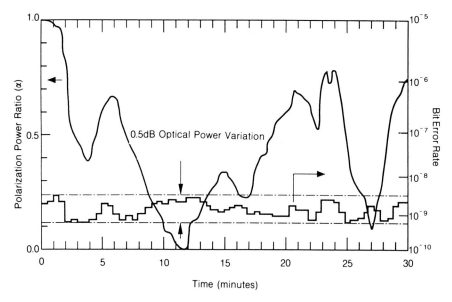

Fig. 7.6 Measured polarization fluctuation and error rate variation.

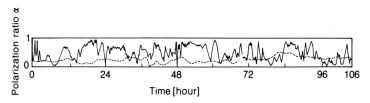

Fig. 7.7 Measured polarization fluctuation.

fluctuation was also characterized by the polarization power ratio. The solid curve was obtained from signals transmitted over the Iyo-Kure link, and changed relatively fast from a polarization power ratio of 0.0 to 1.0. The dotted curve, obtained over the Iyo-Ooita link, changed slowly from 0.0 to 0.5. The difference was mainly due to the mechanical vibration of the fiber in the Iyo-Kure link. The Iyo-Kure link was composed of a longer land cable installed under streets experiencing heavy traffic. On the other hand, the Iyo-Ooita link was composed mostly of submarine cable, where its environment was more stable.

The measured bit error rate performance is shown in Fig. 7.8. The open circles show the performance of the prototype equipment. It was almost independent of the polarization state. The solid squares were obtained under the natural polarization state occurring in the field trial cable. A bit error rate of less than 10^{-15}, which corresponded to error-free operation for longer than 106 hours,

LONG-HAUL TRUNKS AND SUBMARINE COMMUNICATIONS

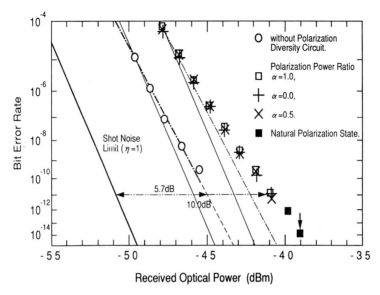

Fig. 7.8 Measured bit error rate.

was obtained. Corresponding observed polarization state fluctuations are shown in Fig. 7.7.

Sensitivity degradation from the shot noise limit, which was calculated by assuming a detector quantum efficiency η of 1, was 10 dB, and that from the best sensitivity reported so far was 4.3 dB. The best sensitivity, 67 photons/bit, was achieved by a receiver without a polarization diversity circuit. The degradation of 4.3 dB was caused by three factors:

(a) insertion loss, 0.8 dB, of an additional optical circuit;
(b) excess noise, 0.4 dB, inherent in the simple addition of two baseband signals demodulated by differential detection;
(c) 3.1 dB due to the imbalance of the two branches and imperfection of the electronic diversity circuits.

If those circuits were carefully and precisely tuned, the degradation could be reduced to less than 1.5 dB. The remaining degradation, that is, 5.7 dB, was caused by five factors:

1. the thermal noise of 0.6 dB;
2. the phase noise of 0.2 dB due to the broad spectral width of the two laser diodes;
3. 1.3 dB due to the 70% quantum efficiency of the PIN photodiode;
4. the front-end optical circuit insertion loss of 0.3 dB;
5. 3.3 dB due to imperfections in the baseband circuits.

In addition to the degradation, the slope of the bit error rate curve obtained by the prototype equipment was gentle compared to that of the shot noise limit curve or the curve measured in a laboratory test.

The target of the optical output power and the receiver sensitivity for a bit error rate of 10^{-11} was 9 dBm and -41 dBm, respectively. The output power target was completely achieved without an optical booster amplifier. On the other hand, the objective receiver sensitivity was only partly achieved. The observed sensitivity ranged from -41.0 to -39.4 dBm. The sensitivity could be easily improved by eliminating electronic circuit imperfections, such as reducing ripples found in the amplitude characteristics of the intermediate frequency amplifier.

Based on these results, a new repeaterless transmission system was developed given in Hayashi, et al. (1991). The new system used a transmitter with a high output optical amplifier in addition to the coherent receiver introduced just before. The output power of 17 dBm was obtained. The achieved system gain of 60 dB gave a repeater spacing of up to 300 km with the use of 1.55 μm low loss fiber and with a system margin of 7.5 dB.

7.1.5 Linewidth and wavelength tuning range of aged DFB-LDs

In order to realize a coherent transmission system, it is also important to confirm the reliability of laser diodes from the viewpoint of coherency characteristics. Linewidth variation (Fukuda, et al., 1992) of a long cavity three-electrode DFB-LD caused by aging is shown in Fig. 7.9. The device was remarkably degraded by threshold current in a forced aging test. The threshold current after the forced aging was 1.6 or 2 times the initial threshold. The increase is estimated to correspond to around 1 000 000 hours of aging in a normal environment.

The rate of increase in the spectral linewidth is relatively slow in the initial degradation stage and becomes more rapid at around the point where the operating current is increasing by 20%. Consequently, if the criterion for life is set to be a 20% operating current increase, the increase in the spectral linewidth can be kept within about 100%. By using this criterion, no problem is expected with the linewidth increase even around a typical operating power of 20 mW at 25 °C over 100 000 hours.

Wavelength tuning range variations of the same devices shown in the previous figure are shown in Fig. 7.10. I_c is the current injected to the center electrode, and I_s means the total current injected to both end electrodes. The wavelength tuning range gradually narrows from large to small current ratio, as degradation becomes severe. The shape of the two curves in the figure is, however, almost the same. Therefore, little difference before and after forced aging can be found in the wavelength tuning range.

In conclusion, the coherency characteristics, for example linewidth, wavelength tuning range, and so on, for a long cavity three-electrode DFB-LD are not expected to fatally degrade within 100 000 hours of aging.

LONG-HAUL TRUNKS AND SUBMARINE COMMUNICATIONS 229

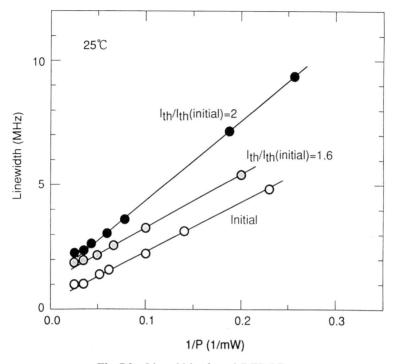

Fig. 7.9 Linewidth of aged DFB-LDs.

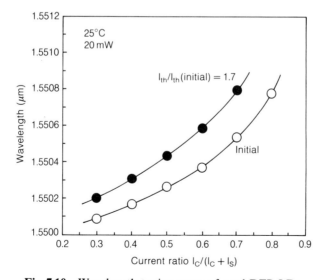

Fig. 7.10 Wavelength tuning range of aged DFB-LDs.

7.2 OPTICAL FDM NETWORKS AND SWITCHING SYSTEMS

Kiyoshi Nosu

In a communications network, where a number of nodes or stations or terminals are connected to each other, highly effective multiplexing/demultiplexing and switching functions are essential. Of the known multiplexing and switching schemes, namely space division multiplexing (SDM) and frequency division multiplexing (FDM), time division multiplexing (TDM) is now widely used in current digital telecommunication systems to create the back bone of the network, i.e. trunk networks. This work is cited in:

Chlamtac and Franta (1990)
Maxemchuk and Zarki (1990)
Acampora and Karol (1989)
Cochrane and Brain (1988)
Hill (1990)

So far, the transport format based upon electric time division multiplexing has been employed because of its ability to easily handle multiplexed data streams and its compact equipment size. However, electrical processing has an inherent limitation with regard to the processing throughput that limits its effectiveness for the era of visual communication. One motivation for developing optical processing systems is to realize the very high throughput systems that are impossible with electric processing. Duplicating the development of electrical signal processing, optical research has concentrated on three multiplexing schemes, namely optical TDM, optical SDM and optical FDM (see Fig. 7.11). The comparison of the three schemes is summarized in Table 7.3. Of these schemes, optical FDM may best improve the flexibility of a communication network. This is the reason why optical frequency division multiplexing and switching are attractive for future telecommunication networks, especially for broadband

Table 7.3 Comparison of optical multiplexing schemes

	Capacity/path or circuit	No. of channels (fiber or line)	No. of paths or circuits/sys.	Flexibility to clock or capacity change
SDM	High	Small ($=1$)	Very large ($>10^4$)	Large
TDM	Medium	Large	Depends on capacity/path or circuit	Limited
FDM	High	Large	Large ($>10^3$)	Large

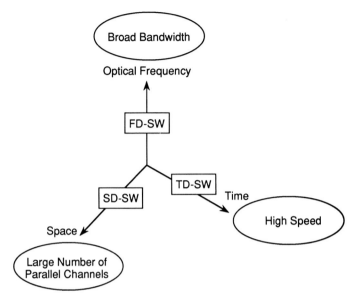

Fig. 7.11 Optical signal processing technologies.

subscriber networks. This work is cited in:

Toba, *et al.* (1990)
Dono, *et al.* (1990)
Prucnal, Santoro and Sehgal (1988)
Hill (1988)
Brain and Cochrane (1988)
Stanley, Hill and Smith (1987)
Hobrinski, *et al.* (1987)
Bachus, *et al.* (1986)
Stern, *et al.* (1987)

One significant difference between trunk line applications and subcriber loop applications is the higher service dependability demanded by the latter. The required transmission capacity of subscriber systems largely depends on the services that the customers are requesting, such as:

- telephone;
- facsimile;
- high speed data;
- picture phone.

Moreover, customer requirements frequently change due to customer relocation, environment changes and terminal changes.

The flexibility needed to respond to these requirements comes from the following advantages of optical FDM:

1. Optical FDM path/channels are independent of transmission speed and format.
2. The communication capacity can be easily increased by introducing optical FDM path/channels.

Broadband information distribution is an attractive application of optical FDM technology. The system configuration of optical FDM information distribution is similar to that of the current coaxial cable CATV, where electric analog signals are frequency division multiplexed and distributed to customers. A tuner at the customer's premises selects one signal from among the signals distributed. In an optical FDM distribution system, optical signals having slightly different optical frequencies are multiplexed at the head end facility and selected by an optical tuner at the user's facility.

There are two ways of separating multiplexed optical signals. One is optical filtering, the other is electrical filtering using optical heterodyne detection. One optical filtering method is to use simple passive components and direct detection. However, this requires narrow pass band filters with a high free spectrum range. Grating filters, Fabry-Perot etalons, waveguide periodic filters (Mach-Zehnder interferometers), array filters, and ring resonators have been proposed and demonstrated for such a method. Electrical IF-band filtering, using optical heterodyne detection, has the advantage of high sensitivity over direct detection and is favored to multiplex a number of channels and to distribute signals to

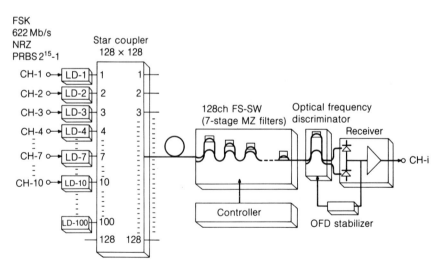

Fig. 7.12 110-channel optical FDM information distribution experiment using optical frequency discrimination detection.
Source: Toba, *et al.*, 1990.

many receivers. However, the heterodyne detection scheme needs local laser diodes with a wide tunable range or complicated circuits, such as an image rejection receiver, to select one of a huge number of optical frequencies.

Figure 7.12 shows the configuration of a 100-channel FDM transmission/distribution experiment conducted at 622 Mbit/s as in Toba, et al. (1990). A polarization insensitive waveguide frequency selection switch with 10 GHz intervals and FSK/direct detection using a periodic (Mach-Zehnder) filter were employed. The wavelengths of the 100 DFB laser diodes were around 1.55 μm and spaced at 10 GHz intervals. The spectral linewidth of the laser diodes was less than 20 MHz. The laser diodes were frequency modulated at 622 Mbit/s with a frequency deviation of 2 GHz. The lightwaves were multiplexed and distributed by a 128 × 128 star coupler. After signal transmission through a 50 km 1.3 μm zero dispersion fiber, the 128-channel tunable waveguide frequency selection switch selected one of the distributed channels.

The channel selection filter consists of seven serially-connected periodic filters

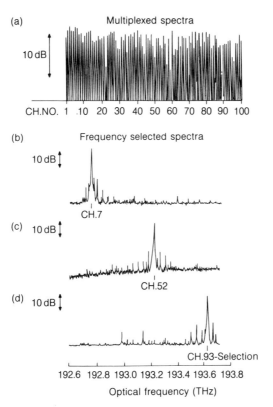

Fig. 7.13 10 GHz-spaced 100-channel optical FDM spectrum and frequency selected spectra.
Source: Toba, et al., 1990.

(Mach-Zehnder filers) with frequency spacings of 10, 20, 40, 80, 160, 320 and 640 GHz. The filter was fabricated from low loss silica waveguides. Frequency tuning was realized by loading thin film heaters onto the waveguides to utilize the thermo-optic effect. The selected FSK signal is converted into an ASK signal by the optical frequency discriminator, which consisted of a Mach-Zehnder filter with a frequency spacing of 2 GHz. The ASK signals from the two output ports of the optical frequency discriminator were differentially detected by the balanced receiver. The multiplexed and frequency selected spectra are shown in Fig. 7.13. Figure 7.13(a) shows the multiplexed spectrum. Figures 7.13(b)–(d) show the frequency selection switch output spectra.

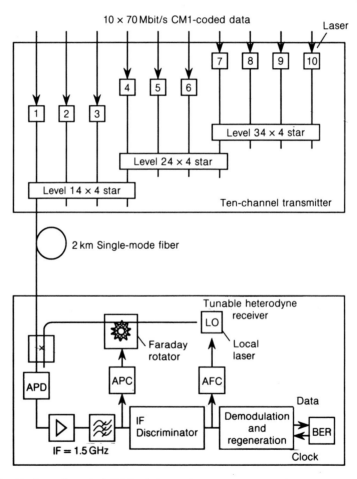

Fig. 7.14 10-channel optical FDM information distribution experiment using optical heterodyne detection.
Source: Bachus, *et al.*, 1986.

OPTICAL FDM NETWORKS AND SWITCHING SYSTEMS

Another channel selection is to use heterodyne detection. Figure 7.14 shows an example of tunable heterodyne detection. The basic configuration is the same as Fig. 7.12. Ten lasers having different optical frequencies are combined by a multiport star coupler. The tunability is realized by changing the frequency of the local laser. The basic principle of heterodyne detection is similar to that of a conventional radio receiver.

Another promising application area of optical FDM is multiaccess networks using optical frequency self-routing. Figure 7.15 is an example of optical FDM channel selection recorded in Arthurs, *et al.* (1988). In this system, optical signals from all nodes/terminals are combined and distributed to all nodes through a center star coupler. The physical network configuration is a simple star configuration, but a virtual mesh configuration is realized by the optical FDM circuit. This system consists of two tunable multiwavelength optical networks, a transport network and a control network. The transport network is formed from tunable wavelength transmitters and fixed wavelength filter receivers, while the control network consists of fixed wavelength transmitters and tunable wavelength filter receivers. Packets at an input port are stored temporarily in a buffer until a request-to-send signal is received. The request-to-send signals are broadcast over the control network to all input ports. At an input port, a wavelength

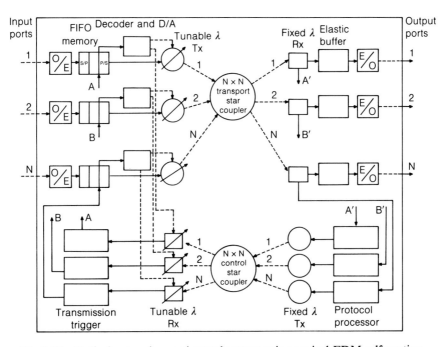

Fig. 7.15 Optical network experimental system using optical FDM self-routing.
Source: Arthurs, *et al.*, 1988.

tunable receiver selects the control information from its desired output port (the output port to which the input port is trying to send information). The packets are routed and transmitted to the appropriate output port by tuning the input port laser to the unique fixed wavelength address corresponding to the output port.

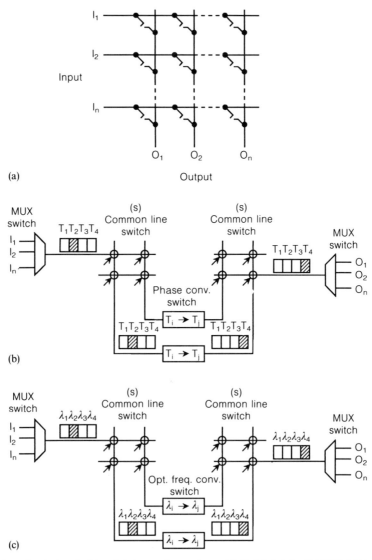

Fig. 7.16 Optical switching fabric configurations. (a) SDM fabric, (b) TDM fabric (STS type) and (c) FDM fabric (SFS type).

7.3 OPTICAL SDM AND TDM NETWORKS AND SWITCHING

Kiyoshi Nosu

In section 7.2 we have discussed optical FDM network systems. The configuration discussed could be applied to switching fabrics for switch nodes as in Kobrinski (1987), Chlamtac, Ganz and Karmi (1989), Goodman, *et al.* (1988) and Arthurs, *et al.* (1988). Other schemes, optical TDM and SDM schemes, would be possible. Therefore, we would like to start with the different switching fabrics possible for switching nodes. Figure 7.16 shows the three basic configurations of switching fabrics: SDM type, TDM type and FDM type. Among them, optical space division switching will also play an important role in broadband transport networks, since it constitutes a switching fabric that preserves optical carrier frequencies, transmission speed, as well as transport format transparency.

These transparency features will push SDM switching to intra-frame connections and intra-office switching systems. Figure 7.17 shows an experimental 8 × 8 optical matrix switch system for a future broadband subscriber network described in Matsunaga, *et al.* (1990). The polarization-independent silica waveguide SDM matrix switch was used to provide multimedia services including 64 kbit/s G4 facsimile, 10 Mbit/s LAN interconnection, 32 Mbit/s NTSC TV and 400 Mbit/s HDTV. On a single mode fiber line, 1.29 µm down stream information and 1.3 µm up stream information were wavelength division multiplexed and the 8 × 8 matrix switch controlled the information streams.

The basic optical TDM switch architecture is the same as the electric TDM switch architecture according to Suzuki (1986) and Matsunaga and Ikeda (1985). It has the potential to construct a switching fabric that is very compact.

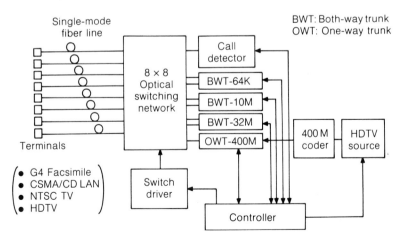

Fig. 7.17 Optical SD switching experiment.
Source: Matsunaga, *et al.*, 1990.

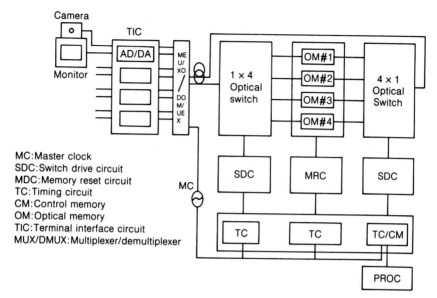

Fig. 7.18 Optical TD switching experiment.
Source: Suzuki, *et al.*, 1986.

Its feasibility largely depends on the performance of optical memories used in the phase conversion switch, where optical pulse positions are interchanged. Figure 7.18 shows the configuration of an experimental optical TDM switch fabric described by Suzuki, *et al.* (1986). Optical bistable laser diodes were used as high speed memories. 64 Mbit/s video signals are multiplexed, converted to 512 Mbit/s optical signals and fed into the optical TDM fabric. In the fabric, a 1 × 4 optical switch demultiplexes and successively distributes demultiplexed optical pulses to bistable LDs. All bistable LDs are reset at every frame period. A 4 × 1 optical switch processes the successively put-up outputs of the bistable LDs.

Before the commercial use of optical TDM switches, we must develop optical memories and logic devices that are faster than eletrical devices as well as optical timing and frame synchronization circuits.

REFERENCES

Acampora, A. S. and Karol, M. I. (1989) An overview of lightwave packet networks, *IEEE Networks*, **3/1**, January.

Arthurs, E. A., *et al.* (1988) HYPASS: an optoelectronics hybrid packet switching system, *IEEE Journal Selected Areas Communications*, **6**, pp. 1500–1510.

Bachus, E. J., *et al.* (1986) Ten-channel coherent optical fiber transmission, *Electronics Letters*, **22**, pp. 1002–1003, September.

REFERENCES

Brain, M. C. and Cochrane, P. (1988) Wavelength-routined optical networks using coherent transmission, *Proceedings of ICC'88*, Philadelphia, PA.

Chlamtac, I. and Franta, W. R. (1990) Rationale, directions and issues surrounding high speed networks, *Proceedings of the IEEE*, **78/1**, pp. 94–120, January.

Chlamtac, I., Ganz, A. and Karmi, G. (1989) Purely optical networks for terabit communications, *Proceedings of INFOCOM'89*, Ottawa, pp. 887–896.

Cochrane, P. and Brain, M. C. (1988) Future optical fiber transmission technology and networks, *IEEE Communications*, pp. 45–60, November.

Dono, N. R., et al. (1990) A wavelength division multiple access network for computer communication, *IEEE Journal of Selected Area Communications*, **8/6**, pp. 983–984, August.

Fukuda, M., Kano, F., Kurosaki, T. and Yoshida, J. (1992a) Spectral aspect of degradation in 1.55 mm long cavity MQW DFB lasers, *IEEE Photonics Technology Letters*, **4**, pp. 305–307.

Fukuda, M., Kano, F., Kurosaki, T. and Yoshida, J. (1992b) Reliability and degradation behavior of highly coherent 1.55 mm long-cavity multiple quantum well (MQW) DFB lasers, *IEEE/OSA Journal of Lightwave Technology*, **10**, pp. 1097–1104.

Gabla, P. M., Leclerc, E., Marcerou, J. E. and Hervo, J. (1992) *Technical Digest of Optical Fiber Communication Conference*, **ThK3**, p. 245.

Goodman, M. S., et al. (1988) Demonstration of fast wavelength tuning for a high performance packet switch, *Proceedings ECOC' 88*, Brighton, pp. 255–258.

Hayashi, Y. (1991) A fully-engineered coherent optical trunk transmission system, *Technical Digest of IOOC'91*, **WeA6-4**, Paris, pp. 393–396, September.

Hayashi, Y., Ohkawa, N., Fushimi, H. and Yanai, D. (1991) *Conference Record of 17th European Conference on Optical Communication ECOC' 91/8th International Conference on Integrated Optics and Optical Fibre Communication IOOC'91*, **WeA6-4**, pp. 393–396.

Hill, G. R. (1988) A wavelength routing approach to optical communications networks, *Proceedings 7th IEEE INFOCOM*, New Orleans, LA, pp. 354–362, March.

Hill, R. (1990) Wavelength domain optical network techniques, *Proceedings of the IEEE*, **78/1**, pp. 121–132, January.

Hobrinski, H., et al. (1987) Demonstration of high capacity in the LAMBDANET architecture: a multiwavelength optical network, *Electronics Letters*, **23**, pp. 824–826, July.

Imai, T., Ohkawa, Y., Ichihashi, Y., Sugie, T. and Ito, T. (1990) *Electronics Letters*, **26/2**, pp. 357–358.

Iwashita, K. and Norimatsu, S. (1991) *Conference Record of the 17th European Conference on Optical Communication ECOC'91/8th International Conference on Integrated Optics and Optical Fibre Communication IOOC'91*, **TcH10-6**, pp. 661–664.

Kahn, J. (1989) 1 Gbit/s PSK homodyne transmission system using phase-locked semiconductor lasers, *IEEE Photonics Technology Letters*, **1/10**, pp. 340–342.

Kahn, J., Gnauck, A., Veselka, J., Korotsky, S. and Kasper, B. (1990) *Technical Digest of Optical Fiber Communication Conference*, **PD 10**.

Kobrinski, H. (1987) Crossconnection of wavelength-division-multiplexed high speed channels, *Electronics Letters*, **23**, pp. 974–976, August.

Matsunaga, M., et al. (1990) Experimental photonic multimedia switching system using integrated 8 × 8 silica-based guided-wave crossbar switch, *Globecom '90*, **706B.2**, San Diego, December.

Matsunaga, T. and Ikeda, H. (1985) Experimental application of LD switch modules to 256 Mbit/s optical time division switching, *Electronics Letters*, **21/20**, p. 945.

Maxemchuk, N. F. and El Zarki, M. (1990) Routing and flow control in high speed wide area networks, *Proceedings of the IEEE*, **78/1**, pp. 204–221, January.

Prucnal, P. R., Santoro, M. A. and Sehgal, S. K. (1988) Ultrafast all-optical synchronous multiple-access fiber networks, *IEEE Journal Selected Area Communications*, **6/7**, August.

Saito, T., Aoki, Y., Fukagia, K., Ishikawa, S. and Fujita, S. (1992) *Technical Digest of Optical Fiber Communication Conference*, **ThD1**, pp. 206–207.

Saito, T., Sonuhara, Y., Fukagai, K., Ishikawa, S., Henmi, N., Fugjita, S. and Aoki, Y. (1991) *Technical Digest of Optical Fiber Communication Conference*, **PD 14**, pp. 65–68.

Stanley, W., Hill, G. R. and Smith, D. W. (1987) The application of coherent optical techniques to wide-band networks, *IEEE/OSA Journal Lightwave Technology*, **LT-5**, pp. 439–451, April.

Stern, J. R., *et al.* (1987) Passive optical local networks for telephony applications and beyond, *Electronics Letters*, **23**, pp. 1255–1257, November.

Sugie, T. (1991) *IEEE/OSA Journal of Lightwave Technology*, **9/9**, pp. 1145–1155.

Suzuki, S., *et al.* (1986) An experiment on hi-speed optical time-division switching, *IEEE/OSA Journal Lightwave Technology*, **LT-4/7**, p. 894, July.

Suzuki, S., *et al.* (1989) HDTV photonic space-division switching system using 8×8 polarization independent $LiNbO_3$ matrix switches, *Technical Digest Photonic Switching Topical Meeting*, pp. 168–170, March.

Toba, H., *et al.* (1990) 100-channel optical FDM transmission/distribution at 622 Mb/s over 50 km, *Proceedings of OFC' 90*, San Francisco, **PD-1**.

8
Epilogue

Kiyoshi Nosu

Key issues of coherent lightwave communication technologies, from coherent detection to optical frequency-division multiplexing, have been described. For the epilogue of this book, we would like to discuss the impact of coherent lightwave technologies on telecommunication networks for the future as proposed by Brackett (1990), Nosu and Toba (1992) and Caruso (1990).

Firstly, we would like to review the present situation of telecommunication networks in Caruso (1990). Telecommunication networks are now no longer just a transport system for telephony, but the infrastructure of the information network society. ISDN is now being used to construct multimedia communication networks mainly for business use. However, recent developments in information processing technologies, such as multimedia personal computers, virtual reality and groupware, will require a huge increase in communication capacity. Figure 8.1 shows the current telecommunication network architecture from the viewpoint of network functions. It has a layered structure for reliability as well as allowing it to expand at low cost. The layered structure consists of the following two fundamental layers:

1. The intelligent layer which is responsible for the control of enhanced service and network operations.
2. The transport layer which is responsible for the connection and transmission of user information.

The transport network constitutes a circuit network, a path network and a transmission media network. The circuit network, which is the highest network in the transport layer, handles the connection of circuits, while the path network manages the bundles of circuits. The transmission media network provides the physical communication link. The fiber optic communication technologies have already introduced the transmission media network in the transport layer. The recent development of optical amplifiers will encourage the construction of bit-rate flexible point-to-point transmission systems. However, due to the commercialization of large capacity transmission systems, the cost of a transmission media network has been reduced drastically, and the relative costs of path and

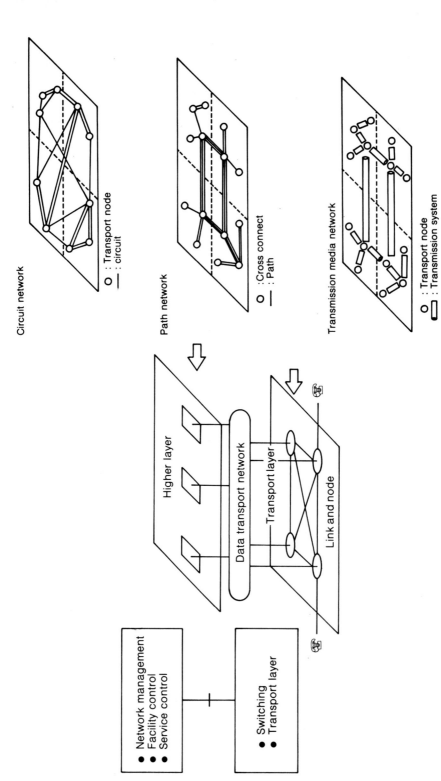

Fig. 8.1 Current network architecture showing hierarchy of telecommunication functions.

circuit networks have become comparatively high. This is only a rough sketch of the current telecommunication networks.

This impact of coherent lightwave communication technologies can be summarized as three fundamental research directions:

- longer repeater spacing;
- higher transmission speed;
- optical multiplexing.

Higher transmission speeds will obviously contribute to the construction of broadband communication systems. Accompanied by the use of optical amplifiers, longer repeater spacings will reduce the dependence of long haul transmission communication costs on distance, even for broadband communication networks. In addition, optical frequency division multiplexing (optical FDM) will create the most fundamental changes in communication networks. Path networks are especially important, because the major part of the path node would be constructed with passive optical devices.

Optical FDM can be effectively used over the extremely broad lightwave bandwidth of 10 000–200 000 GHz, and can realize a novel transport system that would rival current digital (time division multiplexing based) transport networks.

The available bandwidth in the fiber low loss region is about 12 500 GHz (from 1.5 μm to 1.6 μm in wavelength). More than 1000 optical channels are possible within this broad bandwidth, if the channel spacing is set at 10 GHz. If this huge number of optical channels is realized, optical FDM will be used not only for simple large capacity point-to-point transmission but also for novel signal processing technologies at a transport node as an alternative to time division multiplexing, which is the basis of current digital networks. Table. 8.1 summarizes major optical FDM functions:

1. Simple FDM multiplexing combines a number of optical circuits and paths

Table 8.1 Comparison of optical multiplexing schemes

	Capacity/path or circuit	No. of paths or circuits/fiber	No. of paths or circuits/sys.	Flexibility to clock freq. and system throughput
Optical TDM	Medium?	Large	Depends on capacity/path or circuit	Limited
Optical SDM	High	1	Very large (>10 000?)	Large
Optical FDM	High	Large	Large (>1000?)	Large

that are transferred at different carrier frequencies. The function is similar to bit or cell multiplexing in the TDM scheme.
2. Channel selection filters out a specific optical frequency channel. This corresponds to time slot selection in the TDM scheme.
3. Adding (dropping) makes specific optical frequency channels filter in (out) and also allows other channels to pass through. This corresponds to time slot drop/insertion in the TDM scheme.
4. Optical frequency interchange is realized by optical frequency conversion, and corresponds to time slot interchange in the TDM scheme.

The above functions are primarily realized by optical passive devices so that their function is independent of the bit rate, bit synchronization, or transport scheme of information flow. In other words, an optical FDM-based lightwave network is also independent of the bit rate, bit synchronization or transport scheme of information flow. This results in a flexible and expandable communication network. We will be able to change communication capacity without intermediate node rearrangement. From the viewpoint of economics, the intermediate node cost will be reduced because it mostly consists of optical passive devices, while the terminal cost, especially the cost of optical interfaces having a frequency stabilized lightwave, will be increased. However, the cost of coherent lasers will be reduced soon so that the overall system installation cost will be minimized. Figure 8.2 compares the various functions of optical FDM systems.

One of the key technologies for the FDM network is the controllability of optical frequency. The major parameters are illustrated in Fig. 8.3. So far, up to 100 channel multiplexing, demultiplexing and channel selection experiments have been carried out as discussed in the previous chapter. Optical multiplexing is

	System configuration	Items assigned to optical freq.	Applications	Absolute frequency accuracy	Key Devices		
					Filter	Tunable LD	Freq. conv.
Simple multiplexing	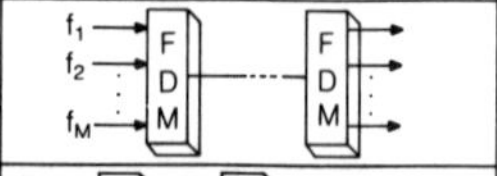	Path and circuit	• Trunk Lines	Moderate		○	
Broadcasting and channel selection	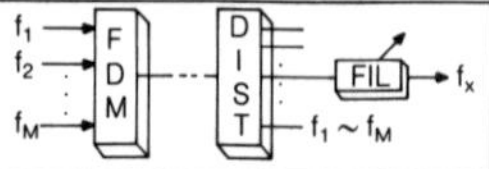	Distributed circuit (program)	• CATV • LAN	Moderate		○	○
Add/drop MUX (taping)	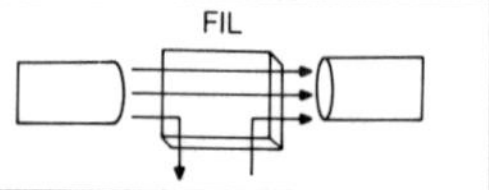	Path or circuit	• Trunk Network • LAN • Inter processor connection	Severe		○	△
Frequency exchange	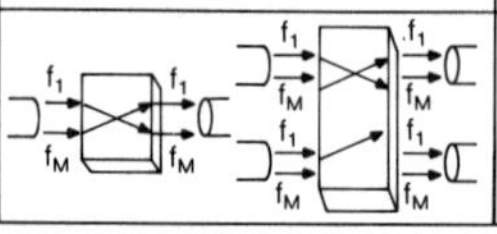	Path or circuit	• Cross-connect /switch	Severe		○	○

Fig. 8.2 The functions of optical FDM. ○ – very important key device; △ – important device.

EPILOGUE

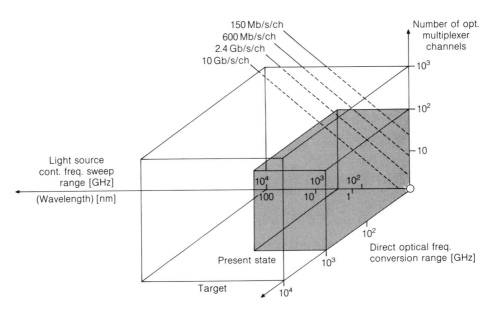

Fig. 8.3 Optical frequency controllability.

almost independent of the transmission capacity of each channel. However, the required ranges of optical frequency sweep and optical frequency conversion depend on transmission capacity per channel because the channel spacing depends on the transmission capacity.

In order to realize a wide area passive optical device-based network that is almost independent of the transmission capacity, the precise evaluation of multichannel optical amplifiers, which simultaneously amplify a number of optical FDM waves, is very important. The performance of common optical amplifiers (discussed in Chapter 6) determines the area size and the number of nodes of an optical network. Roughly speaking, optical transmission performance is a function of the number of optical FDM channels and the optical amplifier gain.

The concept of optical networks is going to be realized in the transmission media network and, consequently, the path network. For example, in metropolitan area trunk networks, the node cost is fairly large in comparison to the transmission line cost. In these areas, traffic density is heavy and traffic flows frequently change due to the reconstruction of buildings, towns and other urban city facilities.

Figure 8.4 sketches structures possible for FDM metropolitan networks. The physical configuration is a simple star configuration or a ring configuration, while all nodes are connected to each other by optical FDM paths. Thus, the "semi-" logical network is a mesh network. The interface of the optical highway and a node is an optical filter and other passive optical devices, all of which are reliable. When the capacity of a path connecting specific nodes must be increased, the

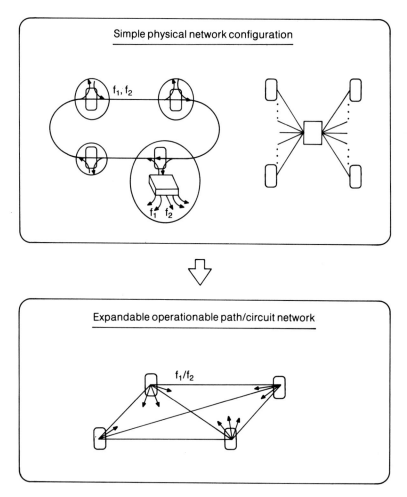

Fig. 8.4 Performance of optically amplified transmission: features of optical FDM-based networks.

replacement of optical transmitters and receivers at those nodes will satisfy the new demand without affecting the other paths or nodes.

The current multiplexing schemes as well as future multiplexing schemes possibly employed for broadband ISDN are illustrated in Fig. 8.5. The optical path contains a number of small/medium size electrical TDM paths. However, the path network for regional trunk systems provides limited transport functions to limited areas. Therefore, optical signals in the network are terminated at the gateways of the network.

Our final goal is the so-called "all-optical" network, which transports optical information without electrical conversion. This includes the three networks of

EPILOGUE

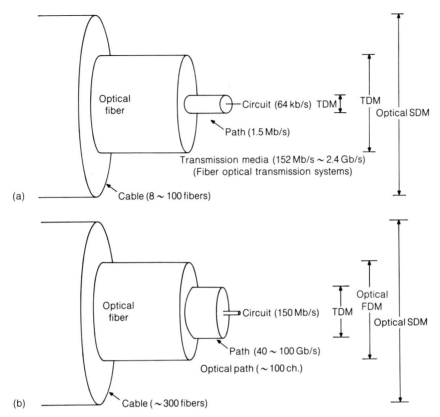

Fig. 8.5 Possible structures of optical FDM metropolitan networks: (a) current multiplexing scheme in the transport layer: (b) multiplexing scheme of trunk line networks in the b-ISDN era.

the transport layer shown in Fig. 8.1. This would be the target for the optical communication networks in the twenty-first century. Figure 8.6 summarizes the possible transition scenario of optical networks, which started from the all-optical link systems using optical amplifiers. The first optimization is being carried out at the transmission media network level. Optical amplifiers realize bit-rate flexible point-to-point transmission, since the bandwidth of an optical amplifier is about 1 THz. The first step should be called "opticalization" of a transmission media network. The next step would be the optical path networks. One way to achieve the optical path networks is the introduction of optical FDM path. The optical FDM technology will offer flexibility in terms of bit rate, bit synchronization, transport scheme (e.g. ATM or STM) of information flows. It will play a significant role in future broadband multimedia communication systems. The ultimate goal is end-to-end optic networks employing optical local switching systems. At this stage, optical time division multiplexing, space division multi-

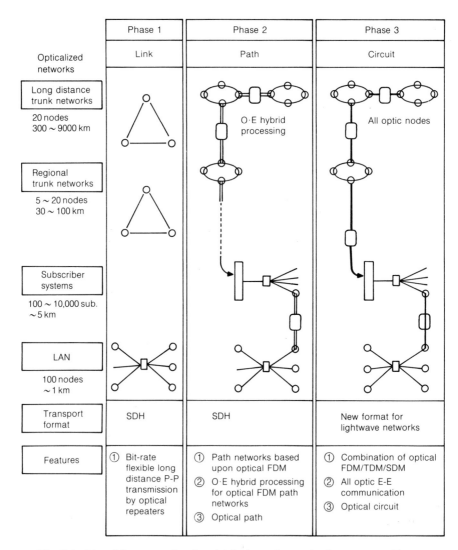

Fig. 8.6 Transition scenario of multiplexing schemes in the transport layers.

plexing, as well as optical frequency division, will be fully used in the transport networks. It will be realized in the twenty-first century.

We have discussed the state-of-the-art coherent lightwave technologies and the perspectives on future lightwave communication networks. When these technologies are cost-effective against customer's demands, multimedia communication (especially virtual reality), groupware and other intelligent visual services will be widely spread consumer activities.

REFERENCES

Brackett, C. A. (1990) Dense wavelength division multiplexing networks: principles and applications, *IEEE Journal of Selected Areas of Communications*, **8/6** (special issue on dense wavelength division multiplexing techniques for high capacity and multiple access communication systems), pp. 948–964, August.

Caruso, R. E. (1990) Network management: a tutorial overview, *IEEE Communications Magazine*, pp. 20–25, March.

Nosu, K. and Toba, H. (1990) Optical FDM-based lightwave network, *Proceedings of ISSSE '92*, September.

Index

Acousto-optic effect 46
Acousto-optic (AO) filters 111, 112
Adhesives in optical paths 202–3
Aging effects
  amplifiers 202
  distributed Bragg reflectors 228–9
Allan variance 130
All-pass filter type delay equalizers 86
Amplified spontaneous emission (ASE) 197
  spectrum 203, 204
Amplifier noise 213–14
Amplifiers, optical 189
  applications 245–7
  common 203–14
  in-line 189–203
  repeaterless systems 221
Amplitude modulators 46
Amplitude shift keying (ASK)
  channel selective receivers 162, 163, 169, 171
  frequency division multiplexing 234
  interchannel crosstalk 205, 210
  linewidth influence 26
  optical heterodyne detection 14, 16–19
    electrical field 14
    and FSK system 21
  power spectrum 39
  receiver sensitivity 35, 36, 37, 38
  repeaterless systems 220
Arc lamps 2, 4
Arc oscillators 5
Arrayed waveguide grating filters 104, 105
Atomic reference 132–5

Balanced heterodyne receivers 76–9
Balanced phase-locked loops

homodyne detection 92
synchronous detection 35
Bandpass filters (BPFs) 13–14
  FSK system 20
Beam Propagation Method (BPM) 123, 124
Beat pulse frequency stabilization 140
Binary phase shift keying (BPSK) 38, 39
Birefringence, polarization dispersion 54–5
Black body sources 1
Bragg grating filters 104–5
Brillouin gain profile 62–3
Brillouin scattering, see Stimulated Brillouin scattering
Brillouin spectral width 62
Brillouin threshold 61
Broadband information distribution 232
Bulk-optic couplers 120–21
Bulk waveplates 89, 90

Calibration, laser frequency 136
Capacitive peaking 75
Carrier effects 47
Carrier effect tuning 135
Carrier wave (CW) power 63–5
Cascaded amplifier systems 189–203
Cascaded MZ filters 109, 110
  channel selective receivers 175–6
Cesium (Cs) atomic clock 136
Channel selective receivers
  heterodyne detection 156–68
  optical filter and direct detection 168–82
Channel selectivity 10
Channel spacing reduction 162–4
Chemical vapor deposition (CVD) 114–15

INDEX

Chromatic dispersion 49–54, 58
  crosstalk 147, 151
  delay equalization 85–7
  transmission characteristics 83, 84
Coherent detection 9–10
Coherent length 1–2
Coherent light 1–5
Coherent optical detection, *see* Optical coherent detection
Coherent time 1–2
Continuous phase frequency shift keying (CPFSK)
  amplifiers
    fiber dispersion limit 193
    noise accumulation limit 194–5
    nonlinear effect limit 197
    performance 198, 199
  differential detection 28–31
  frequency division multiplexing systems 147
  power spectrum 39–40
  receiver sensitivity 36, 38, 218
  repeaterless transmission system 221
  stimulated Brillouin scattering 63–5
Corrugation pitch lasers 45
Coupled equalizers 85
Couplers, optical 120–21
  fiber-optic 121–2
  integrated-optic waveguide 122–5
Cross-phase modulation (XPM) 65–7, 72–4
  crosstalk 144, 147, 149–50
Crosstalk
  amplifiers 213
    fiber amplifiers 210–11
    LD amplifiers 205–7
  channel selective receivers 159–60, 163–4, 175–82
  four-wave mixing 67, 70
  frequency division multiplexing systems 144–56
Current address method 168

Damped oscillations 3
Damping coefficient 3
Decision circuits (DECs) 16, 17
Decision-directed phase-locked loops
  homodyne detection 92
  propagation time effect 96
  receiver sensitivity 38
  synchronous detection 32–5
Delay equalization 83–8
Demodulation 13
  optical heterodyne detection 15
  receiver sensitivity 36–7
Demodulators 13–14
Differential detection
  receiver sensitivity 36, 37
  see also Continuous phase frequency shift keying; Differential phase shift keying
Differential phase shift keying (DPSK)
  amplifiers
    fiber dispersion limit 193
    nonlinear effect limit 197
  channel selective receivers 163
  linewidth influence 26–8
    and continuous phase frequency shift keying 30
  optical homodyne detection 22–3
  receiver sensitivity 35, 36, 38
Diffraction grating filters 104, 105
Direct detection channel selective receivers 168–82
Direct frequency modulation of laser diodes 47–9
Directional coupler type filters 103–4
Distributed Bragg reflector (DFB) laser diodes
  aging effects 228–9
  direct frequency modulation 48
  frequency stabilization 135
  linewidth 43
  modifications 45, 47
  see also Multisection DBR lasers; Three-section DBR lasers
Distributed feedback (DFB) laser diodes
  direct frequency modulation 48–9
  frequency
    fluctuations 130–31
    stabilization 135
  linewidth 43
  modifications 45, 47
  see also Multi-electrode DFB lasers; Multisection DFB lasers tunable 111, 112–13, 117

INDEX

Dithering, laser frequency 135
Dual-shape core-index profile 53–4

Electrical field 13–14
Electrical noise 16
Electrical signal processing 10
Electromagnetic waves
  research history 5–8
  spectrum 1, 2
Electro-optic effect
  modulators 46
  polarization control 89
Elliptical core deformation 55
Envelope detection
  linewidth influence 24–6
  optical heterodyne detection
    ASK system 16, 17
    FSK system 20
  receiver sensitivity 35–8
Erbium-doped fiber amplifiers (EDFAs)
  error rate 200
  fiber amplifiers 212
  performance 195–6
  preamplifiers 217–18
Erbium-doped fiber optical preamplifiers 10
Error probability
  linewidth influence
    differential detection 27, 28, 29
    envelope detection 25, 26
    synchronous detection 31–2
  optical heterodyne detection
    ASK system 18–19
    DPSK system 23
    FSK system 20, 21
    PSK system 21, 22
  optical homodyne detection 24
  PLL propagation time effect 96
  receiver sensitivity 36–7
Error rate performance, amplifiers 199–202
Error-signal generation 135
Extended-cavity lasers 131, 133
External cavity configuration 45
External reference 132–5

Fabry-Perot interferometer (FP) filters 104
  channel selective receivers 169–70
  transmittance 108
  tunable 110–11
Fabry-Perot laser diodes 43
Feedback stabilization 131–5
Feedback technique 135
Fiber amplifiers 208–12
Fiber dispersion 49–58
  nonlinear effect limit 197, 198
Fiber disperision limit, amplifier systems 193–4
Fiber Fabry-Perot (FFP) filters, tunable 111, 112
  channel selective receivers 170
Fiber nonlinearities 58–74
  in optical FDM systems 144–56
Fiber-optic couplers 120–22
Fiber optic transmission
  preamplifiers, erbium-doped 10
  research 7
Fiber-type polarization controllers 89–90
Filters, optical 103
  configuration 103–6
  frequency division multiplexing 232
  tunable 109–13
  waveguide
    design 106–9
    Mach-Zehnder interferometer-type 109–20
Flame hydrolysis deposition (FHD) 114–15
Four-wave mixing (FWM) 65–72
  crosstalk 144–56, 147, 150–56
  LD amplifiers 206–7
Free spectral range (FSR)
  Fabry-Perot filters 169
  ring resonators 108, 109
Frequency counters 137–8
Frequency deviation dependence 174–5, 178
Frequency division multiplexing (FDM), optical
  amplifiers 203–14
  applications 9, 10
  channel selective receiver 156–68
    optical filter and direct detection 168–82
  fiber nonlinear effects 144–56
  four-wave mixing 67

Frequency division multiplexing *cont'd*
  frequency stabilization and measurement 129–38
  future 243–5
  image rejection mixing 82
  multichannel frequency stabilization 138–44
  networks 230–36
  optical coherent detection 13, 38–9, 40
Frequency (phase) noise
  linewidth influence 43, 44
    differential detection 27, 28, 31
    synchronous detection 32, 33, 34
  optical coherent detection 24
Frequency selectivity 13
Frequency shift keying (FSK)
  channel selective receivers 162, 163, 169–73, 176–82
  fiber dispersion
    chromatic 58
    polarization 56–8
  four-wave mixing 154–6
  frequency division multiplexing 233–4
  LD amplifiers 207–8
  linewidth influence 25
  optical heterodyne detection 14, 19–21
    electrical field 14
    waveform 15
  power spectrum 39–40
  receiver sensitivity 35, 36, 38–9
  repeaterless systems 220
  see also Continuous phase frequency shift keying
Frequency spacing dependence 177–8
Frequency-to-output voltage conversion 28, 29
Fusion/tapering method, fiber-optic couplers 121
Future of coherent lightwave technologies 8–10

Gain spectrum
  fiber amplifiers 208–9
  LD amplifiers 203, 204
Gaussian noise 16–17
GeO_2-doped core/silica cladding
  material dispersion 50, 52
  stimulated Brillouin scattering 62, 63
Group delay difference 56

Half waveplates 89, 90
Halogen lamps 2, 4
Heterodyne detection, optical 13–23, 74–83
  amplifier performance 190
    noise accumulation limit 194–5
  AT&T experiment 6, 7
  channel selective receiver 156–68
  frequency division multiplexing 234–5
  phase shift keying 31–2
  receiver sensibility 35, 36, 38–9, 218
History, electromagnetic wave research 5–8
Homodyne detection, optical 23–4, 92–6
  amplifier performance 190
  phase shift keying 31, 32
  receiver sensitivity 35, 37, 38–9, 218
  sensitivity 10

Image rejection receivers (IRRs) 82–3
  channel spacing reduction 162–4
Incoherent light 2–3
Inductor peaking MMIC configuration 75
In-line amplifier systems 189–203
Integrated-optic couplers 120–21, 122–5
Intelligent layer 241
Interchannel crosstalk
  fiber amplifiers 210–11
  LD amplifiers 205–7
Interference thin-film filters 104
Intermediate frequency (IF)
  delay equalization 85
  filters
    channel selective receivers 157–9
    frequency division multiplexing 232–3
  linewidth influence 24, 25, 26
  optical heterodyne detection
    ASK system 17
    power 16
    PSK system 21
    waveform 15
  phase diversity 79–80
  polarization diversity 91
  receiver sensitivity 38
ISDN 241, 246–7

Kerr effect 65–74
  amplifiers 197
  modulators 46

INDEX

Laser diodes (LDs)
  advantages 43
  amplifiers 203–8
  direct frequency modulation 47–9
  emission spectrum 8
  linewidths 24, 43–5
Laser frequency
  fluctuations 129–31
  measurement 136–8
  stabilization 131–6
Lasers
  coherent 8
  coherent time 1
  invention 6
  types 43
Light-emitting diodes (LEDs) 8
Lightwave frequency synthesizer 135–6
$LiNbO_3$ modulators 46
Linewidth
  laser diodes 43–5
    aged distributed Bragg reflectors 228–9
  channel selective receivers 173–4
  and optical coherent detection 24–35
Liquid crystal Fabry-Perot (LCFP) filters 110–11, 112
Long repeater spacing systems 13, 38
Long-span transmission systems 217–20
Low pass filters (LPFs) 13–14

Mach-Zehnder interferometer (MZ) filters 103–4
  cascaded 109, 110, 175
  channel selective receivers 169, 170–71, 173, 175, 179–82
  characteristics 117–18
  configuration 113–14
  design 106, 107, 108–9
  frequency division multiplexing 233–4
  multichannel frequency stabilization 139
  multiple 118–20
  silica waveguides 114–17
  tunable 109–10, 111, 112
Magneto-optic effect 46
Marconi, Guglielmo 5
Material dispersion 49–50
Mercury vapour lamps 1
Metropolitan networks, FDM 245–7
Microstrip line delay equalizers 86–7

Microwave oscillators 24
Modulation indices
  differential detection 30, 31
  envelope detection 25, 26
Modulators 46–9
Molecular reference 132–4
Monochrometers 137
Monolithic microwave IC 77–8
Monolithic tunable lasers 164–5
Multichannel frequency stabilization 138–44
Multi-electric DFB lasers 45, 47–9
  PLL propagation time effect 96
Multimedia 241
Multiple-beam interference filters 103, 104, 105–6
Multiple phase shift DFB lasers 45
Multisection DBR lasers 164–5
Multisection DFB lasers 164–5
Multisection interferometric lasers 165
Mutual gain saturation
  fiber amplifiers 209–10
  LD amplifiers 203–5

Negative frequency feedback 45
Noise accumulation limit 194–6, 198, 199
Nonlinear effect limit 197–8, 199
Nonsynchronous detection 91–2

On-off keying (OOK) 14, 16–19
Optical coherent detection 13
  heterodyne 13–23
  homodyne 23–4
  linewidth influence 24–35
  power spectrum 39–40
  receiver sensitivity comparison 35–9
Optical frequency discriminators (OFDs), detuning 173–4
Optical sources 43–5
Optical spectrum analyzers 137
Optogalvanic signal detection 135

Parabolic-index profile 53, 54
Performance, amplifier systems 190–92
  affecting factors 202–3
  error rate 199–202
  expected 198–9
Phase diversity
  heterodyne receivers 79–82
  receiver sensitivity 37

INDEX

Phase-error variance
  PLL propagation time effect 93
  synchronous detection 33–4
Phase-locked loops (PLLs)
  propagation effect 92–6
  see also Balanced phase-locked loops;
    Decision-directed phase-locked
    loops
Phase matching
  cross-phase modulation 67
  four-wave mixing 67, 70
  optical homodyne detection 23
  self-phase modulation 67
  stimulated scattering processes 67
Phase modulation 197
Phase modulators 46–7
Phase noise, see Frequency noise
Phase shift keying (PSK)
  amplifiers 197
  channel selective receivers 162, 163
  cross-phase modulation 73, 149
  linewidth influence 31–5
  optical heterodyne detection 14, 21–2
    electrical field 14
  optical homodyne detection 92, 95–6
  power spectrum 39, 40
  receiver sensitivity 36, 37, 38, 218
  repeaterless systems 220
  self-phase modulation 72
  see also Binary phase shift keying;
    Differential phase shift keying;
    Quadrature phase shift keying
Phase tunable DFB lasers 47
Photodiodes (PDs)
  channel selective receivers 171
  multichannel frequency stabilization 142
  optical heterodyne detection 15
    ASK system 16
    balanced receivers 77–9
PIN photodiodes 77–9
Pockels effect 46
Point-to-multipoint FDM transmission
    systems 146–7
Point-to-point FDM transmission systems
    146
Polarization
  compensation 88–92
  control 89–90
  dispersion 54–8
  diversity 89, 90–92
    channel selective receivers 164
    repeaterless transmission system
      221–2
  fluctuation 225–8
  scrambling 89
Polishing method, fiber-optic couplers 121
Power spectra 39–40
Power spectral density (PSD)
  channel selective receivers 159
  laser frequency fluctuations 130, 131
Probability density function (PDF)
  linewidth influence
    differential detection 27
    envelope detection 25
    synchronous detection 31
  optical heterodyne detection
    ASK system 17
    FSK system 20
    PSK system 22
Propagation delay difference 56
Pump wave 59
  simulated Brillouin scattering 61
  stimulated Raman scattering 59–60

Quadrature phase shift keying (QPSK) 36, 37
Quantum noise
  linewidth influence 27, 28
  optical heterodyne detection 16
Quarter waveplates 89, 90

Radio wave oscillatiors 5–7
Raman scattering, see Stimulated Raman
    scattering
Raman threshold 60
Random access optical heterodyne receivers
    167–8
Rayleigh scattering 218
Reactive ion beam etching (RIBE) 115
Reactive ion etching (RIA) 115
Receiver sensitivity 13
  channel selective receivers 160, 161
  comparison 35–9
  four-wave mixing 151, 153
  LD amplifiers 207–8
  linewidth influence

INDEX 257

differential detection 27–8, 30
  envelope detection 26
  long-haul trunks 227–8
  optical heterodyne detection 13
    ASK system 19
    thermal noise 16
  optical homodyne detection 24
  phase diversity 80
  polarization diversity 91
  relative intensity noise 76
  short noise 217–18
  thermal noise 74
Reference pulse frequency stabilization 140, 142–4
Relative index difference 50, 52–3
Relative intensity noise (RIN) 76
Repeaterless systems 217–25
Resonance filters 103, 104–5
Ring resonators 104, 107–9
  multichannel frequency stabilization 141–2

Scanning interferometer frequency stabilization 140–41, 144
Segmented-core index profile 53–4
Selectivity, frequency 13
Self-phase modulation (SPM) 65–7, 72–4
Self-routing, optical FDM 235–6
Sensitivity, see Receiver sensitivity
Shot noise
  linewidth influence
    differential detection 27
    envelope detection 25
    synchronous detection 33, 34
  optical heterodyne detection 15
    balanced receivers 77
    thermal noise 74
  optical homodyne detection 23, 24
  receiver sensitivity 217
Shot noise limit
  linewidth influence 28
  optical heterodyne detection 16
Sideband locking frequency stabilization 141, 144
Signal gain
  bandwidth 212–13
  LD amplifiers 203–5

Signal-to-noise ratio
  channel selective receivers 176
  error-signal generation 135
  optical heterodyne detection 16
    ASK system 19, 21
    balanced receivers 76, 77
    DPSK system 23
    FSK system 20–1
    PSK system 22
  optical homodyne detection 24
  receiver sensitivity comparison 35
Silica
  Mach-Zehnder interferometers 114–17
  material dispersion 50, 51, 52
  stimulated Brillouin scattering 62, 63
Spacing, amplifier 192–3
Spacing division multiplexing (SDM) 230, 236–7, 243
Spatial hole burning 44, 45
Spatial incoherency 2–3
Spectral linewidth, see Linewidth
Spectrometers 137
Stabilization, optical frequency 129–38
  multichannel 138–44
Step-index profile 52–3, 54
Stimulated Brillouin scattering (SBS) 59, 61–5
  frequency division multiplexing systems 144, 147, 148
  phase matching 67
  repeaterless systems 218–20
Stimulated Raman scattering (SRS) 59–61
  crosstalk 144, 147, 148–9
  phase matching 67
Simulated scattering processes 59–65
  phase matching 67
Stokes wave 59
  stimulated Brillouin scattering 61
  simulated Raman scattering 59–60
Submarine communications 217–29
Sunlight 2, 4
Switching systems 230–38
Synchronous detection
  frequency stabilization 140, 141–2
  linewidth influence 31–5
  optical heterodyne detection 19
  polarization diversity 91
  receiver sensitivity 36, 37

Telecommunication networks 241–8
Temporal incoherence 2–3
Thermal noise 16, 74–5
Thermal tuning 135
Three-section DBR lasers 47
  channel selective receivers 164–5, 166
Time-division multiplexing (TDM) 230, 236, 237–8, 243–4
  and optical FDM 10
Transmission characteristics 83, 84
Transmission distance, repeaterless systems 220–21
Transmittance
  Fabry-Perot interferometer filters 108
  Mach-Zehnder interferometer filters 106, 108, 117–18
  ring resonators 108
Transport layer 241
Transversal filters 104, 105–6
Triangular-index profile 53
Tunable lasers for heterodyne detection 164–5
Tunable optical filters 109–13
Tunable twin-guide (TTG) lasers 165
T-value 52–3
Two-beam interference filters 103–4
Two-sample variance

laser frequency fluctuations 130, 131, 132
laser frequency stabilization 133
Two-section DBR lasers 47

Utilization efficiency 10

Vacuum tube oscillators 6
Vernier effect 109
Voltage control oscillators (VCOs) 22
V-value
  chromatic dispersion 52
  polarization dispersion 55–6

Wave generation efficiency 69–71
Waveguide dispersion 49–54
Waveguide equalizers 85, 86
Waveguide-type optical filters 106–9
Wavelength counters 137, 138
Wavelength tuning range, aged DFB-LDs 228–9
White noise
  linewidth 44
  optical coherent detection 24

Zeeman effect 135